Managing
High-Tech Start-ups

Managing
High-Tech Start-ups

**Duncan MacVicar
and Darwin Throne**

Butterworth-Heinemann
Boston London Oxford Singapore Sydney Toronto Wellington

ISBN 0-7506-9247-2
Library of Congress Catalog Card Number: 92-73693

Butterworth–Heinemann
80 Montvale Avenue
Stoneham, MA 02180

Linacre House, Jordan Hill
Oxford OX2 8DP
United Kingdom

10 9 8 7 6 5 4 3 2 1

Printed in the United States of America

Editorial, design, and production services provided by HighText Publications, Inc., San Diego, California

Table of Contents

Acknowledgments   ix

Prologue   xi

Chapter One: A Different Kind of Business   1
    Where Do I Start?   2
    Starting a Company: Living on the Edge   3
    Unique Product + Unique People = Success   6
    Razors and Computers: Is There Really a Difference?   10

Chapter Two: Planning for Success   15
    Beyond the Next Bench   15
    Objectives, Strategies, Tactics. . . What's the Difference?   17
    Good Strategies Make Good Companies   25
    Positioning: High-Tech Chess   27
    Think of Your Business Plan as a Treasure Map   29
    Stay Loose   32

Chapter Three: Organizing the Company   35
    What's in a Name?   35
    INC. Is Where It's At   37
    Patents, Copyrights, & Trademarks:
        Guarding the Golden Eggs   42
    The Big Differences in OEM Businesses   51
    Help! Getting Outside Assistance   55

Chapter Four: Funding Your Venture 57
    The Upside and Downside of Venture Capital 65
    Partnerships and Joint Ventures 73
    The Appraisal 75
    Planning for Liquidity with an Exit Scenario 77

Chapter Five: Staffing Your Company 81
    Recruiting the Privates 82
    Employee Benefits Are Good for You, Too 87
    Corporate Culture Shouldn't Be an Accident 94

Chapter Six: Marketing, Sales, and Support 97
    Setting a Price is Like Walking a Tightrope 108
    The Product Splash 111
    Who's Gonna Sell This Stuff? 116
    Sell Around the World in Milliseconds 127
    Some Products Are Really Special 137
    Product Support Sells, and Sells, and. . . . 139

Chapter Seven: Growth Through New Product Development 143
    New Product Development Is Your Fountain of Youth 145
    Develop a Strategy 147
    Duplicate the Success 149
    Making It Manufacturable 153

Chapter Eight: Setting Up a Manufacturing Operation 155
    Make it or Buy It? 159
    Getting It Together 160
    Can We Do It? 165
    How Are We Doing? 167
    Electronics Manufacturing in the 1990s 168

Chapter Nine: Quality is Not a Handbook, It's a Philosophy 171
    Quality in Manufacturing 172
    Quality in Engineering 173
    Quality in Marketing 175
    Quality in Sales 175
    Measuring Performance 176

Chapter Ten: Maintaining Financial Control   177
   Are We Making Any Money?   179
   How Healthy Is Your Company?   181
   Cash Is King   191
   Growing the Business   196
   Monitoring Performance   199
   Garbage In, Garbage Out!   200

Chapter Eleven: How to Get Started   205

Appendices
   A: Business Plan Outline   211
   B: The Product Proposal   215
   C: Volume-Price Graph   219
   D: The New Product Marketing Plan   223
   E: Sales Training Seminar   229
   F: How to Plan for International Sales   231
   G: Product Development Program Manual   233
   H: Components of Manufacturing Information Systems   237
   I: Deming's Fourteen Management Principles   243

Glossary   245

Bibliography   249

Index   253

Acknowledgments

The authors owe a debt to numerous people in Silicon Valley whose assistance and influence helped us create this book. The start-up and growth of a high-tech manufacturing company is a complex process that is mastered by few individuals. This book captures not only our own experiences, but also the experiences of those we have known over our 25 years in the business.

Hank Fallek was a co-founder of Handar and a key contributor to the product strategy, product development, marketing, and sales approaches of that company. Without his persistence in many difficult sales situations, there would not have been a Handar to tell stories about in this book.

Al Jones, a former Hewlett-Packard patent attorney, has for years provided guidance in patent and trademark questions. He was a key contributor to the intellectual property section of this book.

Lon Allen, a partner at Hopkins-Carley in San Jose, is a legal advisor to many start-up companies. His experience in the organization and funding of high-tech corporations provided guidance to Handar from its founding to eventual sale.

Jim Katzman, a founder of Tandem Computer, shared many of the company's start-up experiences, and these are reflected in this book. We are indebted to him for loaning us a copy of their first business plan. Tandem is one of the most socially conscious and financially successful companies in Silicon Valley.

Chris Schwafel, a manufacturing consultant with extensive experience in high-tech industries, provided valuable insight into the state of the art in manufacturing and quality assurance practices.

Charlie Trimble, president and founder of Trimble Navigation, shared his experiences in founding a whole new industry. Charlie's perseverance in raising money to create a commercial business using the military Global Positioning Satellite system (GPS) is characteristic of successful entrepreneurs.

Sam Colella, a partner in the venture capital firm of Institutional Venture Partners, was formerly president of Spectra-Physics. He provided us considerable advice on dealing with the venture capital community and other funding sources.

Bob Mortensen has founded two successful Silicon Valley companies, Quanta-Ray and Lightwave Electronics. He generously shared with us his unique bootstrap approach to high-tech start-ups. Look for his touch in several chapters of this book.

Steve Wurzburg is a partner in Rosenbloom Parish & Bacigalupi, San Jose, which provides legal services to many start-up companies. He helped the authors see through the maze of legal detail involved in founding and managing a high-tech company.

Gordon Baty, general partner in Zero Stage Capital Corp., made many valuable comments in his review of the manuscript that greatly enhance the final product.

Many others have influenced our views on how high-tech companies get started and how they should be managed. We can't list them all here. We can only hope that you are as fortunate to receive the level of support that we have enjoyed over the years.

Duncan MacVicar   Darwin Throne

Prologue

This book is about starting and running a high-technology manufacturing company. In it we offer a broad range of advice for both would-be entrepreneurs and the management teams of start-up companies. The book focuses on high-tech hardware, not software or biotechnology, although much of the material is applicable to these related industries.

Starting a company is easy. A few people come up with a good idea, file incorporation papers, throw a little money in a pot, and they're off.

But surviving the challenges of customer requirements, competitive threats, investors' demands, and cash flow problems... now, that's hard. It's a little like the stock market. There are many ways to evaluate potential purchases, and if you have some spare money, you can figure out what stock to buy. But it's further down the road that you encounter the real challenge: when to sell! So this book deals both with starting issues and the more challenging subject of managing your young company to success.

There is a lot involved in starting and running a high-tech company. Entrepreneurs must therefore be skilled at many different tasks. In our experience, most failures of start-ups are attributable to lack of management skills. So, in this book we have tried to cover all the important management issues for a start-up company. The idea is to be able to cover your weaknesses, whatever they may be. For example, we discuss the importance of international sales and how to set it up. And we offer specific procedures for your marketing and product development departments to follow.

We have organized the book by function—sales, manufacturing, accounting, etc. But in so doing we risk understating the high degree of interdependence of the functions, especially in a start-up company. The marketing and product development departments have to cooperate on strategy planning, or the plans will be weak. Manufacturing and accounting have to work closely and effectively with each other to control your inventory. Sales and product support have to forge an intimate relationship in dealing with customers. And so on. So keep this interdependence in mind as you read. There are hints everywhere.

We prefer that you read this book beginning to end, like a novel. But if that's not your style, you can use the table of contents to find those topics which are the most interesting to you, and just browse through it. It's best if you read the first chapter before browsing, though. Certainly we hope that you will keep the book on your shelf for reference throughout the start-up phase of your company. We have organized it specifically for this purpose. When you have a question, it shouldn't be hard to find where we addressed that subject.

Our hope is that this book can help future high-tech entrepreneurs around the world with solid advice on managing young companies well, from that first meeting around the kitchen table through the eventual financial exit of the company's founders. This is what we've set out to do.

Managing
High-Tech Start-ups

CHAPTER 1

A Different Kind of Business

*T*he professional tennis players who make it to the Wimbledon tournament each summer are good examples for the high-tech entrepreneur. These athletes have the unique ability to focus all their energies on their play, shutting out the rest of the world. They win or lose largely based on their ability to concentrate on the game.

Your start-up company must be equally focused on one all-important item—the product. After all, your company at its inception consists of a group of people working long hours for six or seven days a week to develop the new product. You are excited about the technical innovation you're creating. You are also convinced of the intense need for your product out in the world, usually because of related experience. Naturally, you see your major task as finishing the design and ironing out its wrinkles.

But a start-up company is far more than just a product. *It's a business.* You, the entrepreneur, must see yourself as a businessman or businesswoman right from the start. For the biggest obstacles you will face will usually be business issues—not technical issues.

Once upon a time, two entrepreneurs started a company to manufacture an incredibly sophisticated electronic device. (This is a true story.) The product was a technical marvel, incorporating state-of-the-art design in computer graphics, lasers, electro-optics, electromechanical servo loops, and high-speed electronics. The entrepreneurs faced and solved numerous high-speed synchronization problems which had scuttled previous attempts to build similar devices. Although the market was small, potential customers showed considerable enthusiasm. But the founders had little business experience. When it came to raising funds, establishing distribution channels, setting up manufacturing, and other non-technical tasks, they stumbled. Eventually they were forced to relinquish control of their baby to others who could manage the full business. It was either that or declare bankruptcy. They had created a great product which made a genuine contribution to the world, but they never received their payback.

1

In contrast, another Silicon Valley start-up, Lightwave Electronics, survived its start-up phase due to the business orientation of its founder, Bob Mortensen. Surrounded by a team of technical people who were focused on the new products, Bob set about creating a company. He did some market research to determine if the basic product ideas were sound. He wrote a business plan around those ideas and others. He raised the financing needed to seed the company. He taught himself how to use a personal computer (PC) back when PCs were fairly new in order to track forecasts, budgets, overhead, or whatever. He installed a simple accounting system and established relations with a number of outside resources, including legal, accounting, marketing communications, recruiting, subcontractors, and others. Today, Lightwave Electronics is a profitable small company due largely to the diligent non-technical efforts of its founder.

So it can be done. But it often isn't. Silicon Valley is littered with the carcasses of start-up companies whose great products were killed by poor management.

Our purpose in writing this book is to help you, the high-tech entrepreneur, to face and overcome all these nontechnical challenges in your start-up endeavor. This book covers nearly every topic applicable to the management of a new high-tech venture, and can serve as a reference for you throughout the early years of your company's life. Each topic is supported by real-life experiences—good and bad— of high-tech start-up companies. Our hope is that by following the examples of the successful entrepreneurs we have known you will have a greater chance of success.

Where Do I Start?

Since we have established that business issues are usually more crucial than technical ones in determining whether a high-tech start-up company survives, we now need to examine the tasks facing entrepreneurs when launching a company. Here's a list of some of the basic business management tasks that will be required of you, and which we will discuss in this book:

▼ Prepare a set of business strategies. This exercise must be completed early in the start-up process, before fund raising begins.

▼ Write a business plan. This is an essential step in raising funds, and in preparing to run a company.

▼ Secure financing. This step receives a lot of press, and for good reason—it's difficult! Investors asked to risk millions of dollars tend to ask tough questions.

▼ Staff the company. After financing, probably the second most difficult task you will face.

▼ Establish a marketing function. This is a point that many small-company managers miss.

▼ Set up a sales organization. Another challenging task, and one of the most important decisions the entrepreneur will make.

▼ Provide for post-sales support. How often this critical function of high-tech companies is overlooked! We're gratified by the attention that post-sales support has received in recent years.

▼ Set up your manufacturing department. There is good business leverage, even unfair competitive advantage, to be found here.

▼ Create an atmosphere of commitment to quality. Here is an ingredient that is now essential in any high-tech company.

▼ Establish a finance function. Surprisingly, the need for financial reporting in a small company is no less than in larger companies.

▼ Establish control systems. You may think you can wait to do this—but you can't!

Starting a new high-tech company is a lot like building a football team. Remember what the great Vince Lombardi did? He took a team of professional football players back to the basics—blocking, tackling, play execution—before trying anything elaborate. As a result, his teams had the strong foundation needed to support more sophisticated plays later. His teams won far more than they lost. Your company needs a strong foundation every bit as much as the Green Bay Packers did. That foundation is good business management practices. And you will also need extraordinary perseverance.

So we now know what successful entrepreneurs need to do. But what is it like starting your own high-tech company?

Starting a Company: Living on the Edge

"If I knew it was going to be this difficult, I wouldn't have done it!"
—Charlie Trimble, President
Trimble Navigation.

Trimble Navigation, a Silicon Valley company emerged from the 1980s as the leading supplier of Global Positioning Satellite system (GPS) navigation equipment, a market that is projected to reach multibillion dollar proportions during the 1990s. Like Charlie Trimble, many successful high-tech entrepreneurs have had doubts during the formative years of their company. Few will admit to having doubts of their success because they do not believe they will fail. Entrepreneurial engineers have a "can do" attitude—they will make the company work just as they would make a new product work.

Tandem Computer is an example of one of the start-up successes in the computer industry. It is also a good example of the perseverance required to build a successful company. The original business plan for what eventually became Tandem was written by Jim Treybig and Bill Davidow in 1972, and described a "non-stop" computer system that would be much less likely to suffer "crashes" that brought existing computers to a halt. The plan was based on an experiment that Hewlett-Packard had done with two minicomputers to create a fail-safe computing system. Shortly after the plan was written, Digital Equipment Corporation and Data General announced the development of a fail safe system, and Jim and Bill's business plan was torpedoed. Two years passed. In 1974, Jim realized that the computers announced by Data General and Digital had not been brought to the marketplace. He decided to resurrect the 1972 plan. Almost all of the first year was spent on research and development. Twelve hour days and seven day weeks were not uncommon. The first product was not shipped until 18 months after the company was started. While Tandem succeeded, clearly the pace and pressures of the first year and a half of its existence were extreme.

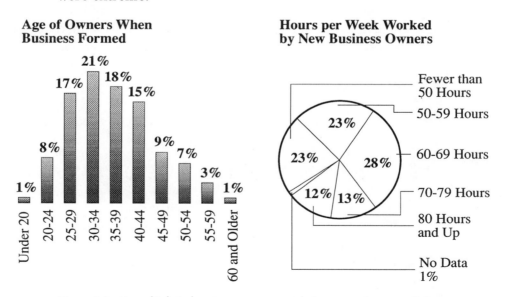

Age of Owners When Business Formed

Hours per Week Worked by New Business Owners

Figure 1-1: Many high-tech entrepreneurs start their companies around the same age that they would normally start families. The enormous amount of time a new company demands often puts a major strain on their personal lives.

Why do people risk their careers, relationships and even health to start businesses like Tandem? Excitement, personal gratification, and "making a difference" are some of the feelings surrounding most start-ups. Recognition for their ideas is another driving force. Ownership of the enterprise and the ability to "make a difference" provide motivation not normally found in large corporations. Contrary to popular assumption, money is seldom the primary motivating factor for most successful high-technology start-ups (although, if the company is indeed successful, money usually follows in ample supply). For many entrepreneurial individuals with a technical background, the most powerful motivation seems to be a desire to have their own "baby" to manage and develop. These people are just as interested in the creative process, the product to be developed, and the creation of the enterprise as they are in the financial rewards. (The investors, of course, are primarily motivated by the financial rewards, but they aren't the ones putting in the 80 hour weeks, eating McDonald's burgers, or often working for very low wages.)

We believe that a commitment to creation of the enterprise and the product is absolutely necessary to sustain the founding management through the difficult start-up years. (Keep in mind that the start-up years in very competitive high-tech businesses can be over a decade and over the $100,000,000 level of sales.) As an example of the commitment required, let's look at the pay entrepreneurs receive. How low is the pay? What personal financial risk do these entrepreneurs take that "normal" people wouldn't consider? Cash is a precious commodity in any company, particularly a start-up. Founders may work for the first few years at one-half to two-thirds of their former pay to conserve cash. Founders and initial employees are often paid the salary that it takes to meet personal expenses. Stock options, otherwise known as "sweat equity," are used to compensate the employees for their sacrifices. Some start-ups that are financed from personal savings do not pay the founders *any* salary until the first product is shipped. Companies that are able to get venture capital financing are not usually quite so austere, but salaries are often lower than in established companies. Ownership in the form of stock and stock options is a significant motivation factor to the founding team and early employees of a company, but more importantly the group must also enjoy the process of building the company. Commitment to the goal is an essential key to the start-up's success.

Besides the immediate financial burden of lower salaries, founders may incur other significant financial risks by pledging personal assets to secure bank loans or other financing. They often leave promising careers in established companies to start the company. The gap in service at the former company may diminish the chances of future success if the business does not succeed and they want to return to the fold. Upon his return, the entrepreneur will have to reestablish his or her position in the fast track.

Nevertheless, the entrepreneurial spirit remains. But the process is full of peaks and valleys. It is not for the faint of heart.

Charlie Trimble's story is illuminating on this point. After several years of modest success selling Loran navigation equipment, Trimble Navigation acquired from Hewlett-Packard the technology to produce a global positioning satellite product. The additional development effort required to bring the new product to market strained the company's resources. However, the technology was good enough to license to a large aerospace manufacturer. An agreement in principle was made in October with a vice-president of one of the divisions of that aerospace firm, but the negotiations dragged on. Over six months passed. The continued development expenses had drained cash reserves to the point that Trimble was on the verge of not making its payroll. With disaster looming, Charlie demonstrated some of the true spirit of the winning high-tech entrepreneur. He told the other firm they had just one week to make up their mind or the deal was off.

Before the week was up, Charlie had deposited $2,000,000 in the corporate bank account. Today, the GPS product line is Trimble Navigation's flagship for the 1990s.

This kind of pressure can require a 150% commitment on the part of the entrepreneurs. Camaraderie, recognition in the business community and business successes for the founders are rewards that do not directly transfer to family or friends. The founders can lose such relationships if time and energy is not set aside to foster them. It is extremely difficult to find time to devote to personal relationships when one is working 80 hours a week, but the loss of family and friends can be a very high price to pay for the company's success. This situation is often made more critical by the fact that many companies are started by entrepreneurs at an age when they are also starting families.

Unique Product + Unique People = Success

High-tech companies often get a lot of attention in the press because of their technical prowess and unique products, but more than technology and unique products are required to build a successful company. Unique people are also needed to implement the company's plans. However, a unique product idea usually comes first.

What is a unique product? It occupies a market niche that is not filled by a competitive product (the minicomputer), is a revolutionary product (the personal computer) or is a variation on an existing product that makes a significant contribution in a market (desktop publishing). It is important to understand where your product fits in this matrix, because your challenges will be different depending on where your product fits.

The minicomputer market was a niche in the late 1960s that grew into a multibillion dollar industry. The minicomputer used then-current electronics technology to bring more computing capability to smaller users. This innovation brought scientific computing out of the corporate mainframe and timesharing environments and placed it at the engineer's fingertips. During the early days of the market, engineers wrote most of the applications programs and assembled the systems. The customers were the systems integrators and the systems maintainers. This situation allowed the new companies to devote most of their scant resources to developing new systems software and hardware products and spend less on sales and service than their mainframe competitors.

The minicomputer was more of an engineering tool than a data processing number cruncher used for record keeping and finance. It was a product line extension from mainframe products being supplied by IBM, Univac, and Control Data. These industry giants did not recognize the trend and let the more entrepreneurial Digital Equipment Corporation, Hewlett-Packard and Data General develop this market. A significant business challenge for these newcomers was to develop technology faster than their mainframe competitors. The reduced customer support provided resources to meet the challenge.

By contrast, the personal computer industry grew out of a technology revolution. The microprocessor development in the early 1970s paved the way for the product. Even industry sages like Robert Noyce did not recognize the significance of this development, dismissing Apple Computer as a market force. Steve Jobs and Steve Wozniak did recognize the power of personal computing and along with others created an open system software and hardware environment. This environment combined with low-cost hardware brought computing to the personal level, as users no longer had to share computing resources. The revolution created new consumer distribution channels as other entrepreneurs created the Businessland and Computerland retail computer stores. Software engineers leaped at the opportunity to develop personal software for the new machines. Within a very short time, a loose coalition of entrepreneurs created a multibillion dollar industry.

Desktop publishing illustrates the opportunity to enhance an existing technology with new product offerings. Contrary to what some believe, Apple Computer did not create desktop publishing—they commercialized it. In 1981, Xerox introduced the STAR system. Like many of today's PCs, it had a mouse, windows, and icon user interfaces. All of this technology was developed at Xerox's Palo Alto Research Center (PARC) back in the 1970s. The STAR system consisted of personal workstation computers linked by a local area network (LAN) to a laser printer. Unfortunately, the system was too expensive, too slow, and lacked significant third party software applications. Eventually, Xerox abandoned the STAR system.

Apple licensed many elements of the STAR system from Xerox. Apple's first attempt to make use of the technology was the ill-fated "Lisa" computer. Like the STAR system, the Lisa had a mouse, windows, and used icons to represent files and system components. The Lisa was not successful, mainly because of its $10,000 price tag and limited performance. However, Apple persevered with the technology and in 1984 introduced the Macintosh and, subsequently, the personal laser printer with Postscript software from Adobe. This hardware and the development of desktop publishing software by Aldus and others created the low-cost personal desktop publishing of the 1990s, over a decade after Xerox had pioneered the technology. Apple's innovations were in the application and marketing of the technology, not in the technology itself.

Two unique products developed in the early 1970s at Hewlett-Packard (H-P) lets us contrast the use of "brute force" technology with innovative thinking to develop unique products. In the "brute force" case, H-P invested over $1,500,000 on the development of a 250-MHz real-time oscilloscope which included many proprietary high-frequency integrated circuits. The product advanced the state of the measurement art and had good customer acceptance. The return factor—profit divided by research and development investment—was 40%. In the other case, H-P introduced the 1601 logic analyzer a few years later. The 1601 used an innovative measurement technique and commercial components to build a new logic troubleshooting device. It had no proprietary integrated circuits and a development cost of just $375,000. However, the return factor was a more-than-respectable 167%.

The moral of this story is that you can take advantage of commercial components and still create innovative products. You may have less proprietary component protection, but you can build a patent portfolio based on innovative block diagrams, circuit designs, and software algorithms.

A unique product is always required to start a new company with high prospects for success, but, don't overlook other areas to establish additional competitive advantage. Quanta-Ray, a Silicon Valley manufacturer of research laser systems, adopted an unusual strategy for sales that gave the company an additional competitive advantage. The founders decided that their customer's highest need, after state-of-the-art research tools, was excellent on-site service. The company established the best field service force in their niche; it was also a top notch sales force. The customer benefit was reduced downtime of the equipment—a big plus in the research business. Quanta-Ray grew to be the number one company in its market niche and was eventually acquired by Spectra-Physics, providing a good profit for the founders.

Having defined your product, how do you implement your plan? Who are the unique people that can help build your company? They are people that will dig in to get the job done even when the job is not defined. They will take charge without having to be told to take charge. You cannot write job descriptions for the kind of effort and performance that is required in the early stages of a young company. Charlie Trimble has said "you have to develop the skill to create miracles to be successful in a start-up. In fact, to survive, you have to perfect the process." It helps if all founders share that attitude. Where do you find people that can meet the challenges of building a business?

Recruiting qualified, dedicated and committed staff is one of the most difficult tasks you will continuously face in building your company. Young entrepreneurs that haven't experienced failure can often overcome the shortcomings of little business experience and succeed through sheer energy, willpower, and intellect, but the development and funding of your company can proceed more smoothly if you have some experienced managers in your founding group. Established high-tech companies provide the largest reservoir of proven business talent and are a logical first stop for recruiting an experienced founding team. If your company is going to grow at 50–100% per year, you should have some member of the founding team with experience to handle the growth. On-the-job learning at high growth rates can be extremely risky. There is a caveat: the sometimes chaotic entrepreneurial environment is not suited to many corporate managers used to having large support structures in place.

Try to recruit members of your team from your peer group or through your network of business associates. It is important to build your founder's team with people that have shared values and mutual respect. Jim Katzman, the founding engineering vice-president of hardware for Tandem summed it up this way: "The values and attitudes of the first fifteen employees of the company set the standard for the next 100." If you have trouble getting qualified people interested in helping you start your company, you may want to re-examine your product idea. If you can't generate interest among potential founding employees, you will also likely have trouble getting investors interested.

After you have identified your product and lined up your founders, you are ready to start organizing the company. But before we get into that, let's take a moment to reflect on the some of the significant differences between high-tech electronics manufacturing start-ups and other manufacturing businesses that may seem more mundane.

Razors and Computers: Is There Really a Difference?

High-tech companies are not the only ones with a lot riding on new product development. All companies—even "low-tech" ones—introducing new products into a competitive market are in the same boat. For example, designing a new everyday item, like a shaving razor, is not an easy task. After all, no two beards, legs, or faces are alike. A razor faces the daunting task of providing the perfect shave by painlessly (and bloodlessly) gliding over the hills and valleys of millions of unique faces and legs, leaving them free of hair. The financial stakes in a new razor can be huge. Gillette invested $200,000,000 to develop the "Sensor," a high-tech safety razor. Hundreds of millions of dollars more were then spent by Gillette to promote the Sensor. What would have happened to Gillette's investment if beards or electric razors suddenly became popular?

Regardless of the product you intend to offer, your high-tech start-up company may have more in common with a company like Gillette than you might think. But there are some critical differences. You can imitate some of the things a successful company like Gillette does, but not all of them.

For your company and Gillette, new product development is an inherently risky process. But it is also the lifeblood of businesses that produce products as diverse as razors and computers. As we've seen in previous examples—particularly the one involving Xerox, Apple, and desktop publishing—new products using new technologies have spawned whole new industries. Are Gillette's challenges with a new razor much different than those you face starting a high-technology company? Yes and no. A significant one for both companies is technology risk:

What if the new razor doesn't work?

Like high-tech start-up companies, Gillette found itself on the leading edge of technology when developing the Sensor. The inherent market risk was compounded by technology risk; Gillette had to develop new plastics and manufacturing technologies. High-tech start-ups run into similar problems all the time. Some, unfortunately, don't cope as well as Gillette. For example, Trilogy was started in the early 1980s in Silicon Valley to create the world's fastest mainframe computer. The technology was so advanced that Trilogy developed its own computer-aided design (CAD) software for integrated circuits because there was no software available that could handle the design complexity. At that time, the CAD industry was in its own start-up mode. Trilogy wound up losing more than $200,000,000 because it could not master the integrated circuit and packaging technology necessary for its computer.

But taking technology risks can also pay off, as the Sensor did for Gillette. 3Com Corporation was incorporated in 1979 to develop personal computer networking technology, which was then in its infancy. The company was launched in the midst of the stormy personal computer industry development in the early 1980s. At the time, there were no accepted domestic or international networking standards, making pioneers extremely vulnerable to design changes in personal computers. Computer communications was compared to the Tower of Babel—a lot of talking but no communication. 3Com staked out its position in the middle of this confusion and took a leadership role in developing the market and the standards for computer networking. They emerged from the 1980s as one of the leading companies in network technology for personal computers.

Regardless of whether your product is a razor or computer, here are some of the big issues you will probably have to deal with when developing and introducing a new product:

▼ rapid technology change

▼ quick market shifts

▼ developing a sales force

▼ convincing your customer of your product's benefit

▼ developing your name recognition

▼ providing post-sales support

Let's take a look at how Gillette and your start-up might handle these issues.

Technology changed at lightning speed in the 1970s and 1980s. The microprocessor was only about two years old when the personal computer was introduced in 1975 by Altair, a New Mexico company. At the end of the 1980s, mainframe computing power was available on a single chip. The average life cycle for many high-technology electronics products is less than three years. Even the car industry has changed dramatically. Japanese car companies redesign cars every three years. The German car industry was working on a 10 to 15 year redesign cycle. In the 1990s, BMW plans to introduce a new model or new engine every year. You must be swift of foot to stay at the forefront in today's high-tech world. You might also be envious of Gillette, because safety razor technology has not changed significantly since its invention over 100 years ago. While the basic technology hasn't changed, however, Gillette is still faced with the task of developing products offering new benefits for its customers.

Markets for any product can change at a rapid rate. The market for citizen's band (CB) radios grew rapidly in the mid-1970s only to collapse in waves of red ink when consumers lost interest in such products. Video games became the hula hoop of the early 1980s as companies like Atari and Coleco grew at dizzying rates. There seemed to be an insatiable demand for the new hardware and software coming out of their development laboratories. Graphics quality and execution speed of the games improved with each new generation of products. But consumers abruptly lost interest in the mid-1980s, and Atari and Coleco were totally unprepared. Today these companies are struggling for survival as scaled-down enterprises pursuing much smaller markets. By contrast, people have been removing body hair through wet shaving for centuries, and this trend is likely to continue.

Gillette can mount a formidable sales effort to assure the success of its new razor. Unlike Gillette, a start-up company seldom has access to a well-established distribution channel to sell its products. Razors are relatively easy to sell; they do not have to be sold by measurement or data processing experts. Many high-technology start-ups require technical expertise beyond the traditional selling skills associated with most sales positions, forcing them to hire specially-qualified sales representatives or create their own sales forces. And finding shelf space for a new personal computer from an unproven vendor at established distributors is very difficult today. The cost to establish a new distribution channel in that market is exorbitant today.

Gillette doesn't have to do much to educate potential customers about its razors, since they have an obvious user benefit—the inexpensive removal of body hair with maximum comfort. Are the features, advantages and benefits of your product as obvious to your customer? The challenge of getting the benefit message out to the customer is often overlooked by high-tech entrepreneurs. The following statement describes a common sentiment: "It's a great product, it will sell itself!" It's never that easy! Many high-tech products have to be sold through an educational process because the product often changes the way customers perform their jobs. This is a paradox of successful new product development: finding a niche that is not currently filled, but having to educate the customer on the product's benefits.

Gillette has brand name recognition; you don't. Unlike Gillette, high-tech start-up companies must establish their reputation and have only limited resources with which to do so. The development of customer relationships is a long, arduous process. For many high-tech products, this is a bigger challenge than the purely technical issues. The development of the personal computer market at the end of the 1970s illustrates this point. The personal computer revolution was created with relatively modest technology—the 8-bit microprocessor. There were many technical hurdles to overcome, such as operating system development, applications software development, fast mass storage, low cost, and fast printers, but these were not overwhelming. Beyond the purely technical issues, the early personal computer entrepreneurs had to show the benefits of personal computing to the world and create a new awareness on how to process information using PCs. Apple Computer selected the famed high-technology public relations firm of Regis McKenna to carry its message to the world. Apple also capitalized on an environment that was very favorable to entrepreneurs to position itself as the symbol of the American dream. According to Apple and Regis McKenna, a new industry was being developed by two young entrepreneurs in a garage and they were having fun doing it. Thus the Apple mystique was born. By contrast, Gillette had it relatively easy with the Sensor. Everyone knows what a razor is and how to use it.

Finally, razors don't require support after the sale. High-technology start-ups must provide product support or risk the loss of repeat or referral business. There are few organizations qualified to service new high-technology products that are not made in high volume production like personal computers. Service policies and organizations must be developed to repair the product when it fails. Few customers are so brave (or foolish) as to spend thousands of dollars on a new product without some assurance that a mechanism exists for answering questions they might have, solving problems in using the product, or repairing it as necessary. Gillette has it easy by comparison. No one ever expects Gillette to repair one of their razors. When one fails, customers simply go buy another.

So is there really a difference between trying to sell someone a razor or a computer? There are numerous factors high-tech companies must be concerned with—such as educating potential customers about the product and abrupt shifts in customer preferences—that a company like Gillette doesn't have to worry about. But your high-tech start-up and Gillette will both prosper only by successfully developing products offering real benefits to your customers. On that pivotal issue, there is no difference between Gillette and your high-tech start-up. For, as we said at the beginning of this chapter, your company is more than just a product. It's a business.

CHAPTER 2

Planning for Success

Beyond the Next Bench

Not long ago, the electronics industry was little more than groups of engineers designing products for their fellow engineers. Raw performance was the measure of the product. This phenomenon, called the "next bench syndrome," drove the industry for decades. But no longer! Today's electronics companies compete based on their ability to satisfy customer needs and not just invent nifty technology. So once you've gathered your unique team and agreed upon the product you plan to build, heed this simple advice: *before you invest lots of time and money on your invention, make sure that somebody out there wants to buy one.* Many companies have failed because they ignored this simple fact.

You may think that you have to design your product before you can discuss it intelligently with others, but that's not true. All that's needed is a simple product drawing and some brief specifications. Granted, sometimes it's necessary to invest in a prototype before making final decisions. But even if you plan test site evaluations later, it's essential to validate your product concept up front. You can also learn much from market research that will help you plan the start-up phase of your company.

Market research produces two key outputs: a definition of the product (we'll discuss this in detail later) and an intimate knowledge of the market. This knowledge is essential for the preparation of your business plan.

How do you gather all this information? There are many sources for the type of information you seek, such as trade magazines and journals, commercial databases, competitive sales literature, industry experts, trade associations, and market research publishers. There are market research firms that can perform focused work for you at reasonable cost (i.e., a few thousand dollars). And some management consultants specialize in helping entrepreneurs through all aspects of market research and product definition. However, the most effective form of market research is talking to potential customers. A few good customer interviews are better than a thousand library sources.

Identifying the people to interview usually isn't too difficult. Your network of personal acquaintances may contain potential customers or at least referrals. Directories of professional societies or industry organizations are useful. You can also look through periodicals where customers are likely to publish. In the worst case, you can call the customer's workplace "cold" and politely talk yourself through the organization to the right person.

When you call potential customers for an appointment, you must describe your purpose in terms of its benefit to them. For example, if your product idea is a workstation with higher speed than any currently available, you could ask an engineering manager's help "to define a new, higher-speed workstation that promises to dramatically improve engineering productivity." Be sure to add that you don't plan to take much of his or her time—certainly less than an hour.

Carry to each interview a questionnaire covering all the issues you need to discuss. This will prevent your forgetting anything. Practice the interview in your mind before preparing the questionnaire, so you can list the issues in a logical sequence.

A few cautions are in order, however. Potential customers will tend to give optimistic answers to your questions, especially in face-to-face interview situations. This means you must test the rationality of their answers with follow-on questions. You also have to keep in mind the economic reality of the customer's situation—can they really afford the wish list of things they're asking for? Finally, market research often produces conflicting results. You will usually have to apply a dose of your own intuition before arriving at conclusions.

No two start-up ventures will have the same market research needs. However, almost all will need information on the items below:

▼ customer needs, both technical (specifications) and motivational (why do they need this product? how badly?)

▼ purchasing influences—who has the authority?

▼ market size and growth

▼ market segmentation by geography, industry, or application

▼ competitors' strengths, weaknesses, and distribution channels

▼ entry barriers, such as government regulations

▼ trends in the marketplace that might affect your business

▼ available distribution channels

Consider the entrepreneur who created a spectacular three-dimensional display for video games in 1985. He put a year of his life—and the lives of several other people—into developing the prototype so he could show it to game manufacturers. Finally, he visited the manufacturers. Imagine his surprise when he learned that nobody would use his display. People liked his invention well enough, but there was scant new product development in the game industry because the bottom had dropped out of the market over a year before. (How's that for timing!) This calamity could have been avoided with a little market research beforehand.

The story of Medical Optics of Carlsbad, California, was very different. The company was founded in 1986 to manufacture ophthalmic surgical systems having less expensive designs than competing systems. The idea was to penetrate the individual office practice with a low price tag. Before the major investor committed its funds, founder Dale Osborne spent about a month talking to a dozen ophthalmologists. The result was positive. He learned that the product concept was valid—ophthalmologists would purchase systems for individual practices at the right price. He identified changes needed to make the product most attractive. He gathered useful information on his planned distribution channel, and he found several good potential customers. Not bad for just a month's work! Soon the company was built, and the product was launched successfully.

There's a hidden benefit to market research like that done by Medical Optics—the people you consult during the research will often become your most loyal customers. After all, the reason you visited them was because you considered them experts, and that's flattering. They may even feel that they have a stake in your company.

Market research need not be expensive. You can scale the research down to just what's required for your situation. For an industrial product aimed at a homogeneous market, visits to six potential customers may be all you'll need. A product with mass appeal needs a broader approach, however, so you may need to hire professional market researchers. Once your market research has verified the direction you want to go, *then* you can begin to plan your venture.

Objectives, Strategies, Tactics…What's the Difference?

Large corporations do a lot of formalized planning. Each year at Hewlett-Packard, the management team prepares a long-term plan, a more comprehensive one to five year plan, and a detailed one year plan. The planning process takes a large chunk of H-P managers' time, but the time is well invested as the managers gain a thorough understanding of what's going on around them. Better yet, they all feel committed to the plans due to their participation.

As an entrepreneur, you may think that you can escape putting forth that kind of effort. Sorry—it isn't so.

Planning in a start-up company need not be as formal as in large companies. Your communication needs are much easier since you have fewer employees, and what you have to say is generally simpler. But you must have a plan for your business, for the same reason that a ship has a rudder: to steer in the direction you want to go.

Now, every entrepreneur has a plan in mind. The question is: Have you written yours down?

Committing your plan to paper has two big benefits. For one, it forces you to think through all the business' issues thoroughly. Secondly, it provides a tool for unambiguous communication from you to employees and investors. And, as a practical matter, most sources of funding will require a complete written plan.

We will describe the preparation of your business plan later in this chapter. But first let's look at the major elements of planning for a high-tech start-up.

The Mission Statement

Any business manager should be able to describe his or her business in one simple, concise sentence. This sentence is termed the *mission statement*. The mission statement for your company follows from your effort to define your uniqueness. For a high-tech manufacturing company, the business statement should be in a form like this: "We manufacture (product) using (technologies) for (customers) to (satisfy applications)."

For example, consider what the mission statement of Sun Microsystems might have been as the company was formed: "We manufacture workstation (one operator per machine) computers using commercially available microprocessors, the UNIX operating system, and innovative architecture (the Stanford University Network) for engineers to solve computation-intensive problems."

This hypothetical mission statement would have served Sun well right up to when they designed a new reduced instruction set computing (RISC) processor, the SPARC, which apparently violated the "commercially available microprocessor" phrase. But the company stuck to its original open architecture commitment and attempted to make the SPARC's design an industry standard rather than keep it proprietary. Now there's consistent business planning!

Be sure to state your mission statement in terms that even the shipping clerk can understand. To check your simplicity, imagine that you're at a cocktail party and someone asks you what your company is all about. Can you answer with an abbreviated mission statement that makes sense? Someone from Sun Microsystems might have answered, "We make engineering workstations."

This mission statement is more than just a documentation of your company's purpose. It's a means by which everyone in the company can have a source of pride as they explain their venture to family and friends. Done well, it will focus everyone on the company's mission.

Your mission statement forms the basis of your business plan, defining product development and marketing priorities and setting the stage for everything that your company will do. For instance, each new product idea should be measured against the mission statement. Does it fit? If not, either drop the idea or rewrite the mission statement (but be sure you have a thorough understanding of the implications of such a change to the company).

Suppose you were planning a start-up to manufacture 3.5-inch disk drives for the PC industry, with uniqueness derived from an innovative design which is one-half the size of any competing product. Providing you believe that all your products will feature small size for the early years of the company, your mission statement might be "We manufacture 3.5-inch disk drives using unique miniature design for PC manufacturers and system assemblers to incorporate into PCs."

Corporate Objectives

Once you have written down your mission statement and all the company founders are in agreement with it, you are prepared to decide upon the company's objectives. These objectives are quantitative statements of the directions in which you intend to move the company. Note our use of the term "quantitative." Objectives must be measurable, since you will be using them to measure your progress as your enterprise develops. Objectives should also state specific time frames whenever possible, since time is an important parameter when measuring whether a start-up company—or any business—is achieving its goals.

Corporate objectives usually fall in one of four categories:

▼ owners' expectations (participation, exit scenario, etc.)

▼ size (sales, growth, employees, etc.)

▼ efficiency (gross profit, earnings, productivity, return on assets, etc.)

▼ market position (share, reputation, etc.)

Other topics occasionally used include quality, employee benefits, management style, and contributions to society. These "softer" topics are becoming more common, even in start-up companies, as owners and managers of industry are more aware of the social, economic, and political implications of the business. One start-up company we encountered had the objective of providing employment for underutilized people in Israel! So think hard about the real reasons why you're starting your company. You may be surprised to find how unique your business idea really is.

How many objectives should a start-up have? There is no simple answer, since the number depends heavily on the nature of the business and its founders. But two objectives are probably too few, and ten are likely too many. Let's establish objectives for our hypothetical disk drive company. For simplicity, we will list just the first two:

1. Achieve 20% share of the PC market for 3.5-inch disk drives by year five.

2. Make at least 15% pre-tax profit consistently in quarterly statements starting in year four.

Both of these objectives are consistent with our mission statement, both are measurable, and both have a time frame associated with them. Whether they are achievable or not depends upon the dynamics of the disk drive market. (We don't pretend to know that; remember, this is a hypothetical case!)

Since objectives are "big picture" statements, there is a temptation to state them in fuzzy terms. In our example, the entrepreneurs might have set objectives like:

"Be the best 3.5-inch disk drive manufacturer in the industry."

"Dominate the 3.5-inch disk drive market."

These statements may sound good, but they're not really useful. What does "best" mean? Does "dominate" refer to market share or to aggressive sales tactics? <u>To have value, objectives must be definitive, clear, measurable, and stated in terms of a specific time frame.</u>

Strategies

In your objectives, you've stated where you want your company to go. The next step is to answer how you are going to get there. The term *strategy* is used in many ways. We use it in a rigorous sense—the single action plan that addresses the "getting there" question. A strategy should be simply stated, confined to a single activity, and related to a specific objective. A set of strategies constitutes a company's strategic plan. You can be sure that your strategies will be reviewed carefully by anyone reading your plan. They capture the essence of your business idea, and they're the most exciting aspect of a new business.

How many strategies should a start-up business plan include? Probably quite a few. Each of your corporate objectives should have a strategy to implement it. In addition, there should be one strategy relating to each functional element of the company, such as product development, marketing, sales, and manufacturing.

Consider our hypothetical disk drive company. It's not hard to think of strategies to support the two corporate objectives we outlined before. The following could be in support of the first objective of achieving 20% market share:

▼ Time the development of the two key products so that the second one can be introduced in year three, one year after the first one. (Why was this strategy selected? Because fast introduction of a second product complementary to the first is often necessary to achieve significant market share.)

▼ Concentrate initial sales efforts on "design-ins" for the emerging market of laptop PCs and in retail outlets for IBM clones, then go after existing original equipment manufacturer (OEM) accounts later. (Here the founders are betting that laptop PCs will constitute a large segment of the market within a few years. They also recognize that PC manufacturers might not change vendors as fast as configure-to-order retail outlets will.)

In support of the second objective of making a 15% pretax profit consistently, our disk drive company might follow these strategies:

▼ Establish, through promotion and evaluation units, a reputation for the finest quality so premium prices can be charged. (Premium prices result in good profit margins, but they must be earned in customers' perceptions.)

▼ Design tooling and automated manufacturing processes to achieve the lowest feasible manufacturing cost consistent with good quality. (Never forget that the other side of the profit equation is low cost!)

Any statement of strategy should be complete and not leave questions of how or why the strategy will succeed. For example, the entrepreneur in our disk drive start-up might have written this as a strategy in support of the first objective: "Grow market share from 10% in year three to 15% in year four and then to 20% in year five." But such a "strategy" is really nothing more than a restatement of the 20% objective. It begs the question of just how the company plans to grow, and we're left without a good answer. Be sure your strategies follow the business situation through to its logical conclusion.

Note that the farther away a statement is from your objectives, the less permanent it is. The above strategies for our hypothetical company might have to change as external forces—customers, competitors, and technologies—change, but the company's objectives remain the same. You should review your strategies at least once a year to determine their suitability to the current situation.

Tactical Plans

Strategies are useless without a set of realistic, specific actions for implementing them. These actions are your *tactics*. Let's look at a few tactics that our hypothetical disk drive company might use to support its strategies. Note that none of these tactics are sufficient by themselves to achieve the strategies they support; other tactics are needed:

▼ During this year, achieve "design-in" with at least two laptop PC manufacturers and sign distribution agreements with at least half of a group of targeted retailers.

▼ Get at least six articles published, half of them by outside authors, on our quality orientation in research and development and manufacturing.

▼ Start full-scale development on drives #2 and #3, and open initial research projects on three remaining new products.

Most of us are familiar with tactical plans, since they are used to set short-term objectives for the organizations in which we have worked. In a well-run company, each individual will have a personal set of plans, typically with 3-, 6-, or 12-month horizons. The tactical plan for the corporation is essentially the sum of all these individual plans. But keep in mind that the source of all these tactical plans, ultimately, is the company's set of strategies.

Objectives

Don't confuse objectives with strategies or tactics. Objectives, strategies, and tactics often get confused with each other. Indeed, the dividing lines between them can often be blurred. That's okay. The important thing is to think issues all the way through when formulating your company's plan for its future—long-term and short-term. The way to do this well is with a methodical sequence of logic leading from objectives to strategies and finally to tactics.

To illustrate these points, let's assume that our organization is a military one and we've been ordered to capture Manhattan without firing a shot. A simple plan for the campaign might look like this:

Objective: Capture Manhattan without firing a shot.

Strategy: Cut them off, and they'll eventually surrender.

Tactics:

1. Destroy all bridges, tunnels, and utilities leading into Manhattan.

2. Blockade airspace and harbors.

3. Wait for them to surrender.

In examining this plan, you might be tempted to think that "cutting them off" (i.e., conducting a siege) is your objective. But this move is a strategy, not an objective. The objective is the conquest of Manhattan. A siege is only one of several possible strategies. We could have chosen to infiltrate the city government and take the city from within. Or we might have decided to begin the siege after conducting an infiltration. Note also that destroying bridges is a tactic, not a strategy. The strategy is the siege, and destroying bridges is but one element of the siege.

You might view your own objectives as less grandiose than the military conquest of Manhattan. But running a start-up company is just as hard, and takes nearly as much planning to succeed.

Example: Tandem Computer

Let us turn to Tandem Computer's business plan, prepared in 1974, to see a good example of how these planning elements work together. The founders of Tandem discovered they had a unique product idea in a computer system designed from the bottom up to be fault-tolerant. They also had a unique parallel architecture in mind. They decided they could sell such a computer system to large corporations, like airlines, to use in on-line transactions.

A unique group of people had been attracted to the start-up. All of the founders had worked in similar roles in the computer industry addressing similar applications, and they had all worked together before at either Hewlett-Packard or Kleiner-Perkins, a venture capital firm. The start-up's unique architecture led to a number of areas of competitive advantage. The computer would be faster than its competitors, yet less expensive. It would be much more easily expandable. And it would be "fail soft" instead of merely redundant in its architecture.

From all these observations, the founders of Tandem decided upon one overriding objective for their enterprise—to achieve and maintain a dominant (30% to 40%) share of the fault-tolerant computer market. They were quite focused in their purpose and did not formulate a second major objective.

Tandem's founders developed four major strategies in pursuit of their objective. It's not difficult to see how each strategy relates to the overriding objective:

▼ First, develop a complete set of hardware and software sufficient to satisfy large customers with modest computing requirements. Later, expand both hardware and software to address more complex requirements.

▼ "Rifle-shoot" large OEM and volume end-user customers with basic initial products for fast penetration of the market. Later, go after customers needing larger systems and single-quantity end users.

▼ Have larger development and marketing programs than any competitor. (This is an unusual strategy, designed to generate and maintain greater market momentum than any other company, including such giants as DEC.)

▼ Insure the availability of peripherals tailored to multiprocessor systems.

Tandem developed appropriate tactics to implement their strategies. Can you match the following five tactics with the strategies listed above?

▼ Approach peripherals manufacturers to discuss their tailoring products for compatibility with Tandem.

▼ In the first year, assemble a product development staff of the size appropriate for a $7,000,000 company.

▼ Assemble a modest marketing/sales team tailored to the penetration of initial targeted customers.

▼ Put "hooks" into the original software to facilitate future software upgrades.

▼ Concentrate research and development efforts for the first two years on key components of the initial basic systems.

(The answers? We've simply reversed the order in which the tactics were presented compared to the strategies.)

Tandem's business plan was simple, yet it contained all the elements of a good start-up plan. Proof of its worthiness lies in Tandem's results. The company eventually went public, and in 1988 sales topped $1,000,000,000. That's all the reality check any plan needs! As we've said before in this chapter, good strategies make good companies.

Good Strategies Make Good Companies

Over half of new companies in the U.S. fail within a few years of their founding, and high-tech companies are no exception. While it's not a good idea to dwell on this negative side of the start-up experience, it is important to understand the most common failure mechanisms so you can avoid them.

We have found that many start-up failures are due to inadequate strategies. Founders of high-tech companies sometimes just don't think their business plans all the way through. We can't offer advice that will cover every situation, but we can share the following four principles to help you develop sound strategies—

1. *Focus on the first product.* Most successful high-tech start-up companies started with a flagship product. Can anyone forget what the first Apple computer looked like? It is important that your start-up company focus nearly all of its energies on product #1, as Apple did, and do all you can to make it successful. If you don't, you may never get the chance to work on product #2. So follow Apple's example—push your first product to its ultimate success and consolidate your hold on your chosen niche. Only then should you attempt to expand.

2. *But beware the woes of the one-product company.* Now that we've hammered home the point of focusing your efforts, we must point out that a one-product company is in a precarious position. Any change in markets or the competitive situation can be devastating. EIP Microwave, a Silicon Valley company, was founded to manufacture microwave frequency counters, in competition with Hewlett-Packard. They did it. EIP's share of the market grew to the point where it was as large as H-P's— about one-third of the market. But EIP was never able to come up with product #2. They built several versions of their counter, but this did not protect them from the eventual shrinking of the market and increased pressure by H-P and other players. To be fair, the company did try hard to get into the related market for network analyzers, but the effort fell far short of

expectations. Ten years after its founding, EIP Microwave was floundering. As you plan the course of your company, be sure to map out one or more possibilities for second products with good sales potential. Concentrate on the first product initially, but be fully prepared to launch the second one when the time is right.

3.   *Focus your marketing efforts.* Your start-up company must concentrate its promotional and sales efforts on a market niche in which you're confident that you can be successful. This niche, your "defensible beachhead," is where your uniqueness translates into significant competitive advantage. In this way, you will maximize your probability of success. Once you have identified your niche, be sure that it is large enough to support your plan. Your sales in the niche will usually have to be shared with competitors. Yet your sales must be sufficient to pay for the investments you have made in product development and the development of the market.

Would you believe that there was a start-up company whose first five product introductions were all in different markets? Each product idea was good, but there was no synergy upon which to build promotional and sales programs. As you might imagine, the company was swamped by marketing and sales issues. They never developed significant sales in any one of the markets. Six years after its founding, the company was betting its future on yet another new market/product niche! (Wish them luck....) Contrast that story to the experience of ILX Lightwave of Bozeman, Montana. This company was founded to manufacture test instruments for manufacturers and users of diode lasers. (Now there's a well-defined niche!) In time they developed a broad line of instruments, each addressing another need of their targeted customers. The company grew from stand-alone instruments to complete test systems. With this focus, ILX Lightwave achieved an enviable record of steady growth, and maintained a sizable share in its niche.

4.   *But beware the one-customer company.* Dependence on a single customer is even more dangerous than the one-product syndrome. High-tech customers are notoriously fickle. Now, we're not going to say that you should turn down the chance to sell large quantities of your output to General Motors, but we will advise conservative planning. Here are two examples: A service company to high-tech industry had great prospects with a single

account. The company's management team borrowed funds to increase output capacity and, while they were at it, to build the sales organization to get even more business in the future. But the new business didn't materialize within the expected time frame. By the time the president figured this out, the bloated organization couldn't cover either payroll or interest payments, and it went bankrupt. In contrast, consider the software start-up that landed a great account, IBM, at the expense of their much larger competitors. Instead of resting on this success, the management team redoubled their efforts to find and penetrate other accounts, so they would not be dependent solely on Big Blue for their success. It was hard work, but also smart business management. The strategy paid off when the IBM business proved to be less than anticipated.

Positioning: High-Tech Chess

One of the most important strategies to create is your marketing strategy, describing how you will approach the customer with products and promotion. This strategy arises from your positioning statement.

Positioning gets a lot of play in high-tech industry, and for good reason. Positioning is important, because it's the means by which your product gets matched with your potential customers.

Simply stated, positioning is creating an image in the customer's mind of how your product uniquely fits his or her needs. When this image is effectively communicated, customers can make informed decisions, and you'll get your share of the market.

A well-formulated positioning message, delivered to the market in effective ways, will cause your targeted customers to realize that they should come to you. Equally important to the start-up company, those potential customers whom you can't satisfy will stay away, dramatically lowering your marketing and sales costs and allowing your organization to focus on that important initial market niche.

To develop your positioning statement, start by talking with potential customers about your product. Ask them how and when they would use it. Ask them how they would compare it to similar items, what they think the price should be, and so on. Once you have a complete understanding of the customer's perception of your product, then you're ready to prepare a positioning statement and the promotional campaigns that flow from it. The best positioning is accomplished by understanding the customer's perception of the product and turning it around and feeding it back to him or her.

Let's look at some examples of positioning. When Compaq introduced its first computer, the world quickly understood what it was—a portable version of the IBM PC—no more, no less. Compaq's positioning was accomplished not only by promotion, but by the product itself. It looked portable—it had a keyboard that folded up into a front cover, and a carrying strap on the side. No one could mistake that this was a portable computer that was compatible with IBM's PC. The customers who needed portability showed up in droves.

A less successful story is that of Saxpy Computer Corporation, a Silicon Valley start-up. In 1987, Saxpy built what might have been the world's fastest computer, capable of performing certain computations at the dizzying speed of 1,000,000,000 operations per second. The company's founders decided that they had invented a less expensive version of the Cray supercomputer, which would be attractive to a certain class of customers, so they promoted the system as an alternative to Cray and other computers to such customers. The problem was that the customers didn't see it that way. They saw the Saxpy invention as an ultra-high-speed "process server," a co-processor that would do special computational tasks better than the host CPU. In other words, they perceived Saxpy's product as a peripheral device, not a computer system. In fact, some of them thought it would be an ideal process server for a Cray supercomputer! In the end, Saxpy was unable to differentiate its product from the array of "super-minicomputers" and "mini-supercomputers" that were flooding the market at that time. The company folded, and poor positioning played a big part.

Positioning is applicable not only to a company, but also to each of its products separately. That is, every product you introduce should have its own positioning strategy. A start-up company should tie the company's positioning to that of its first product. This coordinated positioning might last for quite a while. Compaq continued its positioning as the portable computer company for years as they introduced a number of laptop and portable-looking desktops. More often, however, succeeding products need to have their own identity.

Consider the case of Quantum Design, in San Diego, California. Their first product was a magnetometer based on a superconducting electronic device called a SQUID. The system had the fine sensitivity required to perform the most sophisticated superconductivity research. This was in 1988–1989, when high-temperature superconductivity was one of the hottest research topics on the globe. As a result, the magnetometer achieved instant success. The company's positioning at that point could be stated as "the only reliable supplier of SQUID magnetometers in the world." The company then decided to introduce a second product to expand the market. This system was a considerably less expensive version of the original $100,000 system, having correspondingly lesser performance but the same sensitivity. After considering what the new product had to offer the customer, Quantum Design decided to position it with language like "now every lab can afford SQUID sensitivity." So Quantum Design then had two differentiated products, the first with a performance message reinforcing the "only game in town" position, and the second with a price message, allowing it to compete with the home-made systems used by lower-budget research groups.

Your positioning statement is the cornerstone of your marketing programs. Use it consistently in all the ways your company is presented to the customer. When you talk with the press, industry experts, key customers, and other influential people, be sure to communicate your positioning to them. Include explicit statements of your positioning in advertisements, literature, trade show exhibits, and retail packaging.

Managing your image properly throughout all your promotional campaigns is a tricky business. For this job, you need a marketing communications agency to design your stationery, sales literature, public relations materials, advertising, and sometimes even product appearance. With the professional assistance of these agencies, you can insure that every form of communication between your company and potential customers reinforces the image you want. Many young companies shy away from such agencies because they fear the high price of their services, but in reality these services do not have to cost much. All you need to do is follow some simple guidelines:

▼ Develop a clear positioning concept for your company and each of its products. Your agency will be able to help fine-tune the ideas.

▼ Develop clear objectives for any item to be prepared by the agency—ad campaign, brochure, press release, etc.

▼ Set a realistic budget for each item.

▼ Stay in close touch with the agency, and don't allow them to work in a vacuum.

Think of Your Business Plan as a Treasure Map

What's the first thing you would do before going out in search of buried treasure? Get a map, of course. Treasure hunting today is different from the days of skulls and crossbones, but the need for a map is just as great.

Modern-day treasure hunters have a tough job. They have to do considerable research, sifting through hundreds of references to determine the whereabouts and content of the treasure. Having done this research, they can draw a sketchy map of their course. Then they need to raise money; it can cost millions of dollars to conduct a modern treasure hunt. They need to hire people, rent a ship, purchase specialized equipment, and so on. Despite all this preparation, they often have to contend with rough weather, unfriendly locals, and other hazards of the trade.

We all understand treasure hunting to be risky. Experienced treasure hunters work hard to minimize their risks and thus maximize their chances of success.

Your situation as a high-tech entrepreneur is much the same as the treasure hunter. You must work hard to plan your venture, leaving as little to chance as you can. And you need a map showing how you're going to get to your goal. This map—your *business plan*—should contain accurate answers to three basic questions of business strategy:

▼ Where are we now?

▼ Where do we want to be, and when?

▼ How do we get from here to there?

A business plan is a lot more than just a map. It is a communication vehicle between you and others involved in the venture—founders, investors, current and prospective employees, and outside resources such as your accountant or attorney.

The single most important purpose for the plan is raising funds. Don't be discouraged if you're required to rewrite the plan a number of times to suit the desires of potential investors. It may be your business idea, but it's their money!

Preparing the plan is an exercise in self-discipline. As you work on it, you'll find yourself asking over and over again, "Have I forgotten anything?" Appendix A contains the outline of a business plan. This outline is deliberately sketchy in order to show the most important elements in every plan. Your plan will probably have greater emphasis on some elements of the plan than on others.

Above all else, your business plan must capture the essence of your endeavor. It must emphasize your uniqueness and consequent competitive advantage—in other words, why you will succeed. It must contain clear statements of your mission, objectives, and strategies. If the reader understands these concepts well, then you have a good plan—everything else is just supportive detail.

In your plan, discuss all of the company's functional areas—product development, marketing, sales, manufacturing, and finance. Describe the position of every key employee, even if you have not yet identified a person for the slot. Emphasize the intended sales structure, both domestic and international—a weak point in many high-tech plans.

Your plan must outline all the financial aspects of the business. It is important to be conservative in the financial projections; the odds are that you will need more cash than you think. In fact, it's a good idea to describe contingency plans to deal with common emergencies like a shortfall of cash while ramping up production, a delayed product introduction, or a new competitor. Readers of a business plan are always impressed by an honest discussion of risks.

Your business plan need not read like it's conservative, however. It should be upbeat and positive, while professional in appearance. And it should contain a few "uppers"—possibilities for new markets opening up, or for immensely successful sales of second products. Investors will feel a lot better about risking their money on a start-up if you show them a few possibilities on the upside.

You should try hard to protect the confidentiality of your business plan. After all, it has considerable value to you. Make only the number of copies that you absolutely need. Copies should be numbered, and whenever one changes hands, the transaction should be logged. The document should be labeled "proprietary and confidential," or words to that effect. It's a good idea to ask anyone who reads your plan to sign a nondisclosure statement to protect your property. This rule is not applicable to people close to the company such as your attorney, nor to venture capitalists, who make a practice of not signing nondisclosure statements.

From the looks of some business plans, there are people who think that the longer a plan is, the more attractive it will be. This is not the case. The quality of a business plan is far more important than its quantity. If you were a venture capitalist, digesting 40 or 50 new business plans a week, would you be more impressed by a thick one or a thin one? Your business plan should be short enough to be easily understood, yet complete enough to demonstrate that you've thought the process completely through. Make it only as long as needed to communicate the essentials.

This may seem an obvious point, but your business plan won't do you any good unless your founding team commits to following it. It's amazing how many start-up business plans wind up gathering dust on a shelf once funding is secured. Don't discard your planning efforts. Let your plan be a living guide for the management team during your company's formative years.

It is no sin to revise your business plan when conditions change. There are many variables outside your control which might require a shift in your plans, like competitive moves, shifting customer habits, and rising or falling economies. So you need to review your plan at least once each year. Ferretec, Inc., a Silicon Valley manufacturer of microwave components, had a plan to diversify into the instrument business early in the company's history. But fate dealt Ferretec a different hand. Orders for their new component were far in excess of their original plan, and the management team found themselves too busy to pursue the diversification. So they tabled the diversification for a year or so, while strengthening the core component business. When that phase of the company's growth was in hand, they returned to the original plan. After five years Ferretec was in both businesses, and doing just fine. The lesson here is either follow your plan or change it. Either way, you and those around you will know why you are doing what you're doing.

Entrepreneurs often seek help in preparing their business plans. Help of this nature can be valuable, since start-up founders rarely have knowledge of every functional area. Where can you find help? First, your circle of acquaintances may include entrepreneurs or others with experience themselves. At the least, your acquaintances are a good source of referrals. Next, your accountant, attorney, or banker might be of assistance or be able to refer you to a professional. Other sources of referrals are the editors of trade magazines and business journals, who often know the experts in the industry or the local geographic area. The same is true of local offices of the Chamber of Commerce or the Small Business Administration. Venture capitalists and investment bankers may be willing to assist in plan preparation if they feel strongly about the company's chances of success.

Some management consultants specialize in business plan preparation. As a general rule, the large consulting and accounting firms are not set up to help entrepreneurs with business plans, so you should find a solo practitioner or small consulting company. One form in which you will find these consultants is the "incubator" practice, designed specifically to provide a broad range of services to the entrepreneur at modest cost.

No matter how much outside help you receive, remember that this is your business plan. You may gather a great deal of information and advice from others, but you must write the plan yourself.

Stay Loose

Most start-up companies face competitors who are much larger and stronger. These big companies have arsenals of weaponry to fight off competitors, like cash in the bank, sales organizations, loyal customers, and efficient volume manufacturing. Start-ups do have weapons of their own, such as unique product designs, but these often seem meager compared to the big boys. One weapon of start-up companies, however, has the potential to be awesome: *flexibility*.

When changes occur in the business climate, large companies need time to respond. They have layers of decision-makers and procedures to follow. But small companies can change direction quickly, giving them a real competitive edge over their larger cousins. If your company has a flexible attitude about strategy, you will be able to turn out new products faster than your competitors. You will also be able to modify sales channels or promotional campaigns with greater ease.

You may wonder if a flexible posture negates all that good work that went into the business plan. Not at all! By learning about your business environment during the planning process, you have collected the knowledge base necessary to make rapid corrections when things change. Of course, you should follow your plan until there is compelling reason to change it, but don't be so locked into your goal that you lose sight of important changes affecting your business.

There are many factors which can change your business environment. Let's look at some of the more important ones:

▼ *Technology changes.* The drop in price of semiconductor memory during the 1980s was amazing. Just think of the problem a new company would have if its plan depended on competitors' high manufacturing costs due to their heavy use of RAM!

▼ *Applications changes.* Who would have predicted in 1980 that the personal computer would change the face of the entire computer industry? By 1990, Apple and Microsoft were major corporations, clusters of PCs were replacing mid-sized computers and even mainframes, and the dividing line between engineering workstations and high-end PCs was fuzzy. Clearly, any new company in the computer business will be chasing a moving target.

▼ *Geographical market changes.* Remember how attractive China was for high-tech companies in the 1980s? There were many joint ventures and sales relationships forged between the U.S. and China in the decade. But everything changed in 1989, with China's crackdown on student protesters. Certainly many good business plans were rewritten afterwards. Perhaps similar stories will be told about the fallout from "Europe 1992."

▼ *Competitive moves.* 1990 was an exciting year in the workstation business. IBM launched a new, credible line of computers, H-P consolidated its acquisition of Apollo, and Sun Microsystems entered the low end of the market, competing with high-end PCs. Who a few years earlier would have predicted all that?

▼ *Political changes.* In the early 1980s, U.S. government spending on military hardware grew rapidly, driven by the Reagan Administration. But the buildup slowed dramatically in the middle of the decade without any fanfare. During the Bush Administration, the decline accelerated as the Cold War ground to a halt. These shifts were not predicted by most of us. Start-up companies whose business depended on military funds were caught by surprise.

▼ *Regulatory changes.* Government policy on the regulation of an industry is subject to change, sometimes abruptly. The breakup of AT&T had ramifications which still were not worked out by the early 1990s. Manufacturers of telecommunications hardware have been rewriting their business plans frequently, due solely to the litany of court decisions affecting AT&T and the "baby Bells."

▼ *Business cycle.* You might think that a start-up company would be immune to the ups and downs of the business cycle, but that's not so. Capital spending can take a big downturn in recessions, and most high-tech hardware fits into that category. Interest rates, which directly affect your cost of borrowing or raising money, can swing wildly. They did in 1974, 1980, and 1982. They probably will again. A sudden increase in interest payments would be a most unpleasant surprise to any company with debt.

Internal factors can also cause the start-up company to change strategies. The unexpected loss of a key employee can have a major impact. Investors, especially venture capitalists with majority interests, can press for changes. And then there is the most common factor of all—your own mistakes.

The flexibility to respond to change in your company's environment is so important that you should strive to make it a way of life. It should be a part of the culture of your company. When your employees share a philosophy of flexibility, then you will be able to make decisions quickly and delegate responsibility more effectively.

CHAPTER 3

Organizing the Company

What's in a Name?

*T*he name of your company has considerable impact on the image you will present to the customer. So spend some effort on designing your company's name. It's a worthwhile investment, and changing your corporate name later on gets very expensive.

There are no rules governing how a company should be named. Instead, it's largely a matter of personal preference. However, we have some advice to offer on the matter:

▼ *You should like the name.* You can be sure that the founders of Apple Computer enjoyed their start-up's name. Remember, you will personally be associated with that name, so it's important to you.

▼ *The name should create the right images in people's minds.* Consider the name of Valid Logic Systems. Doesn't that name generate the impression that the company's products will always give you the right ("valid") answers?

▼ *It is equally important that your name not create negative images in the customer's mind.* Fortunately, we can't think of any really bad examples.

▼ *Don't use initials or numbers for your company's name.* People don't remember initials well, perhaps because they can't relate them to an image. Names such as IBM or 3M work only after the company is famous. So during your start-up phase use real words. High-tech start-up names such as TSI, LTX, or PRC definitely lack inspiration.

▼ *The name should not have too many syllables.* Americans prefer to speak or hear two, or maybe three, syllables. Granted, you may need to use more than one word for your company's name. In this case, the rule applies to the first name, or to the logical abbreviation of the name, such as "DEC" for Digital Equipment Corporation. Imagine the problems the founders of International Microelectronics Products must have with customer communications! Compare their situation with the founders of Cypress Semiconductor, who have a two syllable first name which is easily remembered and which lends itself to a nice logo—a cypress tree.

▼ *The name should be easily pronounced and spelled.* Entrepreneurs have created some real tongue twisters in their eagerness to invent high-tech sounding names. Don't you imagine that the employees of Visx had to spell the company's name every time they talked on the phone?

▼ *Be sure that your name doesn't step on someone else's toes.* This can happen easily with abbreviations. To avoid this costly error, do some simple research to see if you might be encroaching on another name or set of initials. A logical starting point is corporate directories and checking the ads in trade magazines for your intended markets. You should also check with your state's department of corporations or similar agency; most states will refuse to grant incorporation to companies whose names are too similar to those of already-existing companies doing business in that state. Finally, your attorney can help you with an out-of-state name search, particularly for trademarked names.

▼ *Your company's name must not be too specific about the business you are establishing.* Often, a start-up company changes its plan after a few years. Such a change is hard enough without the additional burden of changing the company's name, so avoid using a name that might hamper your flexibility. An example of this problem was Niravoice, a Silicon Valley manufacturer of voice transmission equipment. When the company changed its business to local area networks, they had to change the company's name to Netcor.

▼ *Your company's name will be presented to customers verbally, written, and graphically, so be sure that it fits easily into all of these forms.* The name must be easy to spell, easy to pronounce, readily adapted to a logo, and easily fitted into graphics. These last two points are especially difficult challenges for most non-artists. You may find the advice of an experienced graphics designer helpful.

The earlier you decide upon a name for your company, the better. Communication with others, especially sources of funding, is greatly enhanced when they have a name to use consistently in referring to your venture. A name also lends more credibility to your story when you are talking to potential customers, business partners, and vendors.

You may be tempted to invent a brand name for your product. Our advice is: *Don't.* Brand names work well for large corporations, who often need to separate the image of the product from that of the company. You would be pretty tired of the name "Proctor and Gamble" if it were used on every product they make—the grocery store would be full of it. And who can deny that "Scotch tape" is a better name than "3M tape"? The objective of the start-up company is to spread its name around, not hide it. Having two names will only confuse things. The lesson here is similar to that of other chapters of this book: namely, focus on a few things—in this case, a single name—in order to do them well and to get your customers to focus on you. One way to insure this focus is to present them with the same identity every time they hear from you.

INC. Is Where It's At

There are several alternatives for the form of a business: sole proprietorship, partnership, or corporation. We believe the only viable form of business, under the current tax laws, is the corporation, or in some circumstances, a subchapter-S corporation. An "S-corp" receives special treatment from the IRS for federal tax purposes; its legal formation is not different from other corporations. Switching the business form from a partnership or sole proprietorship at a later date may raise questions if outside venture capital financing is sought for the company. The corporation provides the following advantages that the other two business forms cannot provide:

▼ *A modicum of liability protection.* The corporation provides a limited shield for the owners (stockholders) of the company if any lawsuits are brought against the company. State laws allow for some limitation of liability for directors in certain circumstances.

▼ *A higher level of credibility in the business community.* The corporate form is recognized as a more standard form by the high-tech business community.

▼ *Established rules of ownership.* The corporation provides a standard form of ownership by the issuance of common and preferred stock. Ownership is determined by the amount of stock issued to each shareholder. The obligations of the officers and directors are better defined.

▼ *Separation of personal and business financial interests.* The corporation becomes the employer of the founders. This relationship isolates the employees' and business' finances. Be sure to follow the corporate formalities such as regular board meetings with appropriate announcements and minutes to maintain the legal identity of the corporation.

▼ *The only viable form if venture capital is required.* The corporate form is the only viable form if venture capital funds will be invested in the company. This is a familiar form for venture capitalists and is necessary for any public offering. Even if the company is to be sold in the future, it is preferable to sell stock rather than assets, because the purchaser assumes all liabilities, shielding the seller from possible legal action by former customers, employees, and vendors.

An S-corporation may be preferred under some financing situations. If a subchapter-S election is made, the profits and losses of the corporation flow to the investors in proportion to the stock ownership. For example, individual private investors may be interested in taking advantage of start-up losses in the form of personal tax benefits to obtain an early return of part of their investment. Subchapter-S corporations are treated differently for state and federal tax purposes. Another advantage of the S-corporation is that there is no "double taxation" on profits that are distributed to shareholders. (This refers to the fact that profits of a normal corporation are subject to corporate income tax and then again to personal income tax when distributed to stockholders.) However, venture capital firms are rarely interested in a subchapter-S corporation. They generally prefer to capitalize start-up costs and take advantage of the tax deduction for the company when it becomes profitable. Your accountant can help you with the decision to elect to make a subchapter-S filing.

Incorporating a company is typically a relatively inexpensive process. While the exact requirements vary from state to state, there are a number of decisions to be made to incorporate:

▼ *Name of the company.* This may sound trivial, but picking a name that is not taken can sometimes be difficult. Your name will be registered with the state of incorporation as part of the incorporation process when articles of incorporation are filed; a call to your secretary of state's office will usually confirm availability of the name you're considering.

▼ *Class of securities to be issued.* The amount of stock, both common and preferred, to be authorized and issued must be determined. The amount of stock issued is arbitrary, but the class of stock is affected by the types of investment in the company. Venture capitalists and outside investors usually get preferred stock with some additional rights over the common or founder's stock. There is often a price difference as well. How stock is distributed to the founders is also arbitrary and normally negotiated between the founders and related to relative worth and any financial contribution that is made.

▼ *Size of the board of directors.* For many start-ups, the board is usually three to four people with one or two representatives from the outside shareholders and one or two insiders. A slate of candidates is selected by the incorporator and the shareholders elect the directors from this slate.

▼ *Capital structure of the company.* The company can be capitalized with a combination of equity (the amount paid by the investors for the stock that is issued) and debt. The company should be capitalized with a debt-to-equity ratio less than three to one to avoid possible unfavorable rulings from the IRS. The company may be deemed a sole proprietorship or partnership if it is formed with a higher debt ratio using notes or bank guarantees from the founders.

Be sure to choose good legal counsel to handle the incorporation process and to advise your firm for the long-term. You should choose the best legal counsel, specializing in high-tech corporate law, that will take you as a client. The best attorneys are in the business to advise the growth client.

Interview prospective attorneys to determine compatibility of approach to problem solving. The chemistry in the relationship is important, because many issues will require an understanding between your attorney and you that goes beyond a simple business relationship. For example, your level of risk tolerance may be an issue in a litigation proceeding or a contract negotiation. A good personal relationship with your legal counsel can facilitate decision-making in these circumstances.

The best legal counsel may also be the least expensive because of their familiarity with most legal issues for a high-tech start-up. You do not want to pay for someone on the steep part of the learning curve;

the cost can be high both in fees and in mistakes. Discuss the fee structure with the prospective counsel to avoid unpleasant surprises and to establish a working understanding of their billing practices.

You should also adopt a stock option plan for your new company. Stock options are a major source of motivation for new employees in a high-tech start-up company and should be used liberally and wisely. The securities issued under the plan must be registered in the state where the option recipients reside or there must be an exemption from registration. Federal securities laws also apply. The plan should include provisions for stock repurchase under a defined formula if an employee, with vested stock, leaves the company. The goal of stock options is to motivate current employees; disgruntled former employees with a shareholder interest can make your life difficult.

Employment agreements for the key employees should also be drafted at this time. The founders are extremely important to the success of the company and should not be able to easily walk away from the situation. Part of the agreement should be devoted to a stock repurchase agreement, otherwise known as a "buy-sell" agreement.

Board of Directors

The board of directors is an important part of your team and you must choose the candidates carefully. Minimize the number of insiders on the board—two at the most. The outside investors will require board representation to look after their interests. An additional board member is usually chosen that is acceptable to both inside and outside investors. Your attorney can often help select a candidate for this position. The board will be elected by the shareholders from the slate of candidates that is usually proposed by management. Shareholders can also elect members not on the management backed slate within the rules the articles of incorporation.

Even if inside shareholders own 100% of the stock, outside directors should be selected to help manage the business. It is very easy to get bogged down in daily activities and lose sight of the big picture. The board can help you avoid that pitfall. For example, Handar, a Silicon Valley manufacturer of data acquisitions systems, was a closely held company with no outside investors. However, it had two outside board members. One was from the high-tech industry and the other was an entrepreneur familiar with government and finance. The high-tech representative made many contributions toward the management of the manufacturing and the sales operations, helping the company trim inventory costs and develop a more effective sales force.

The financial entrepreneur helped Handar obtain a better banking relationship and made introductions to several high-level politicians in Washington to assist in a protest filing on a government procurement.

When you've selected your board, you should use them for the following tasks:

▼ Advise top management on strategic issues.

▼ Assist in recruiting top management executives.

▼ Provide guidance in developing relationships with other companies.

▼ Establish and approve general corporate policies such as benefit plans, compensation guidelines, and stock options.

The board's sole legal responsibility is to the shareholders. The board members have a fiduciary responsibility to see that the company is managed in a responsible, prudent, and legal manner. Listen to your board members; they will give you the most honest opinions you will hear.

Outside members of the board should have a financial interest in your company in the form of stock options or purchased stock. This interest need not be large; up to one-quarter of one percent is common. You should pay all travel expenses for outside board members to attend board meetings. As the company grows, the outside members should be paid for each meeting.

The legal frequency of board meetings is at least once per year. Some outside investors may insist on meeting once a month during the start-up phase. Later, meetings can be held quarterly with a major annual review.

Director liability is a major issue. Many qualified people will not serve on boards because of this exposure. The liability can be mitigated by purchasing director's and officer's insurance; be sure to watch the exclusions in the policy. Indemnification by the company can help, but it could become worthless if the company gets into financial trouble. Venture capital investors may indemnify an outside board member if the person is highly qualified to make a contribution to the company's success.

Other Professional Help

You should select an accounting firm at the time of incorporation to establish the capital structure and types of stock to be issued for the company. A good regional accounting firm can fulfill your needs in many cases, particularly if you are not seeking financing from venture capitalists or other professional investors. A large nationally known accounting firm such as Price Waterhouse or Arthur Anderson will usually be required to get venture capital funding. Venture capitalists will be looking for liquidity of their investment either through an initial public offering (IPO) or sale of the company. In either case, audited financial statements from a nationally recognized accounting firm are required to make the process go smoothly.

The large accounting firm may also be able to help you find capital and help with negotiations on valuations and structuring of any investment. Be careful that this approach does not burden you with extra financial overhead on your staff in the critical start-up days. Some venture capitalists may want too many financial heavyweights on the payroll in the start-up phase to impress the investment bankers at a public offering. (We'll discuss financing of your company in more detail in our next chapter.)

A competent insurance agent that specializes in business coverage should also be selected. Get appropriate insurance to cover operations, equipment, workers' compensation, and liability, and review the coverage regularly as your company grows. Be sure to include business interruption insurance in the package to protect the business in case of natural disaster or other unforeseen calamities. Properly incorporating and capitalizing your company can provide the shareholders with product liability protection, but you should also carry insurance to protect the company itself.

Patents, Copyrights, & Trademarks: Guarding the Golden Eggs

"Xerox Sues Apple: Personal Computer Industry Shaken"

"Polaroid Sues Eastman Kodak: Billions Lost in Instant Photography Venture"

"Texas Instruments Sues Japanese Companies for IC Patent Infringement"

These are some of the headlines that appeared in financial publications in the late 1980s and early 1990s. What happened? Why did technology lawsuits for patent and copyright infringement become so popular? The foremost reason is the change in how patent litigation is handled in the U.S. courts.

In 1982, Congress created the Court of Appeals for the Federal Circuit (CAFC) to hear all patent, copyright, and trademark case appeals from the district courts as well as the U.S. Patent and Trademark Office. All rulings of the CAFC are now the law of the land. At the close of the 1980s, the courts were upholding patents 80% of the time. These changes have provided the spark to make companies more protective of their intellectual property. Companies are now filing for more patents and copyrights to protect their research and development investments and to provide a shield against potential suits from competitors who might sue to enjoin them from producing a product. Prior to 1982, it was not uncommon for entrepreneurs that had concerns about potential patent infringement to be advised that a way could be found to avoid facing that issue. That is no longer the case.

The changes in the U.S. legal system have also caused more involvement of foreign companies in the patent process. Foreign filings amounted to more than 40% of total patent applications in the U.S. in 1989. This change has to be considered in the start-up of a new business. Care should be taken to be sure that new products do not infringe on existing patents or copyrights because companies are quite aggressive in protecting their patent and intellectual property rights.

Intense worldwide competition for new markets is another reason for the escalation of patent and copyright litigation. Some companies, like Xerox, have made many significant developments in the laboratory but have not converted them to meaningful products. They are now looking to their patent library as a source of new revenues through licenses. Licensing fees that used to be about 1% of revenues are growing to the 3 to 5% range. If you are developing a new product that may possibly infringe on an existing patent or copyright, you should consider licensing the technology from the patent holder. And if you are developing new technology, you should take advantage of the protection afforded by the patent, copyright, and trademark laws.

Patents

Software algorithms and hardware designs can be patented to provide the needed protection. Software algorithms that only provide number-crunching capability cannot be patented. For example, a new way to calculate the net present value of an investment could not be patented, but an algorithm that processed a series of atmospheric measurements to calculate prevailing cloud height and then presented that information visually could be patented. There are three tests that an invention must pass in order to qualify for a patent:

1. The invention must be useful—the "pet rock" failed this test so it could not be patented.

2. The invention must be different—this test is almost always passed.

3. The invention must not be obvious to someone skilled in the art—this is a subjective test and the most difficult one to pass.

KLA, a Santa Clara, California company that manufactures test equipment for the integrated circuit industry, faced that last test when the company applied for a patent on a concrete optics bench. The invention dramatically reduced the cost of manufacturing precision optical benches. However, the patent examiner pulled out concrete park benches as prior art and was not going to award the patent! One of the KLA's arguments was that manufacturers of concrete park benches were in a different field and, consequently, the park bench art was not analogous art; someone in the optics field would not consider design of park benches when designing an optical bench. Furthermore, the precision requirements for an optics bench are much more stringent than for a park bench. These arguments were eventually accepted by the examiner and the patent was awarded.

A patent does not give the business the right to manufacture the product patented, but does give it the right to prevent others from making, using, or selling the product. For example, suppose that the three-legged stool had been patented, and you improved on the design by adding a back and tried to get a patent. Even though you improved on the design, you would not be free to manufacture the three-legged chair because to do so would infringe on the three-legged stool patent. Further, suppose that you improved on the existing design by adding a fourth leg to the stool and tried to get a patent. You would still not be home free, because you would infringe on the three-legged stool patent. You must take something away from the patent claims to not infringe on the claims of a patent.

Filing for patents should be based on business needs to determine if the cost is justified. For example, if the product life is less than two years, a patent would hardly be justified since it takes approximately two years to obtain a patent. You should consider the prior art when filing for a patent to build your case with the patent examiner and to help you draft a better application. A good patent attorney can assist you with patent searches at low cost. There are also commercial patent survey services and many libraries maintain patent records that can be searched. If there is a question on whether or not to file, err on the side of filing. Even if your patent is declined, a good defense is established to thwart future claims against you by your competitors.

You should also be aware that information in the public domain may prevent you from getting a patent on your product. For example, many companies use their technical journals as a way of protecting against patent claims by competitors. IBM uses their technical bulletin as a patent defense. It is published by the patent department and includes all of the work that IBM has elected not to patent. This action precludes a competitor from getting a patent on similar work and preventing IBM from using the technology in its products in the future. U.S. examiners wait one year before relying on the publication; foreign offices use the information immediately.

If a patent is filed and is not likely to be granted, then you can publish defensively through the patent office as a defense strategy against lawsuits by competitors that might subsequently obtain patents on the same or similar work. If this isn't done, and the patent is denied, then the patent office destroys all of the application material, thereby weakening your defense.

There are two patent systems in use today in the world. The United States and the Philippines grant patents to the first inventor; the rest of the world grants to the first applicant. If two people file within a year, then there is an "interference" declared by the Patent Office to determine who was the first inventor. The United States invention is a two-step process: conception of the product, followed by reduction to practice, i.e., a working prototype. Once a patent is granted, you have protection for 17 years, but there are periodic maintenance fees that must be paid to keep the patents in force. Patents are enforceable, even if the infringer did not know that the patent existed.

There are also time limits on filing for a patent. In the United States, you must file within one year from public disclosure of the idea or first sale of a product based on the idea. In the rest of the world, you must file before there has been an "enabling disclosure" of the idea. "Enabling" means shipping a product that can be reverse-engineered or providing documentation on the theory of operation that facilitates the engineering of a similar product. You may want to file early in foreign countries if your idea has great promise for sales in the foreign market. For example, in Japan, very little development need be done to file for a patent—the concept does not have to be reduced to a working prototype as in the United States.

The question of whether or not to file internationally is always a difficult one to answer. International patents are expensive to obtain and difficult to defend. The following rules should help you make that decision:

1. Where will you have manufacturing and volume sales?

2. Where do you project future sales growth?

3. Who are the competitors, where are they located, and what impact will they have on this product?

For example, Philips Corporation, headquartered in the Netherlands, is one of Hewlett-Packard's biggest competitors in Europe in instruments. Approximately 5% of H-P's sales are in the Netherlands, so they only file in the larger European countries. Telebit is a growth company in communications. Third world countries are major growth markets, so they are filing there. International protection for patents and copyrights has been improving, so there are more advantages in filing than there were in past years.

Patents are not a panacea for competitive issues. You receive a monopoly to protect the target market from competition with the same technology, but be aware that if the market is attractive, competitors will try to design around your patents. The development of the video cassette recorder (VCR) illustrates this principle. JVC, a Japanese company, went to Sony, another Japanese company, with a request for a license to use the Betamax VCR technology. Sony refused. After this rebuff, JVC teamed up with Matsushita and designed the VHS system. They proceeded to license the technology to any company interested. VHS has become the industry standard and eventually killed Betamax, forcing Sony to use the VHS technology for its own VCRs! Had Sony licensed the Betamax technology to JVC, Betamax might still be a factor in the VCR market.

Protecting patents can be expensive and large companies are difficult to scare if they want to come after you. An out-of-court settlement can easily cost at least $20,000 to $50,000; full-scale litigation such as the Polaroid versus Eastman Kodak case over instant photography can cost millions of dollars. It is usually best to settle on reasonable licensing fees.

Even with patent protection for your products, your business must still be effective in sales, marketing, and manufacturing. Patents can also be helpful in the financing phase of a start-up because investors, particularly venture capitalists, see them as a potential weapon against competition, independent technical validation of the product, and the source of possible licensing fees or technology exchange.

United States patent applications are not prohibitively expensive to process; it takes about two years to get a patent issued. The words "patent pending" is not required on your product in the meantime. Applications require about 30 to 40 hours to complete, with attorney's fees typically in the $100–$250 per hour range. A good patent attorney will try to help a client broaden the claims on the patent. For example, Al Jones, a Silicon Valley patent attorney, had a client with a patent idea for a toothbrush for handicapped people. The brush handle was expensive to fabricate, so the inventor created a new latch and key mechanism where the key was part of the throw-away brush. After a lot of browbeating, Al convinced the inventor to also patent the latch mechanism. There have been several inquiries about licensing of the latch, but none about the toothbrush.

The patent process is started by filing a patent disclosure with the Patent and Trademark Office. The disclosure establishes the date of conception of the invention and is retained for two years. The disclosure does not obviate the need to diligently proceed with the filing of the patent. The inventor should be aware that public use or sale in the United States or publication of the invention anywhere in the world more than one year prior to filing of a patent application will prohibit the granting of a patent. More information and patent application forms can be obtained from U.S. Department of Commerce, Patent and Trademark Office, Washington, D.C., 20231.

Trade Secrets

Many companies do not file patents on all of their technologies to avoid disclosure of certain processes or company know-how. This activity falls into the realm of trade secrets. For example, the details of a semiconductor process or magnetic head coating process may be carefully guarded. There have been numerous cases of litigation between employer and former employee in Silicon Valley concerning trade secrets. Be aware of the sensitivity of this issue, especially if any of your products will be competitive with those of your former employer or former employers of your employees. One way to avoid litigation is to get an appropriate release from former employers. If you or any of your employees have signed nondisclosure or trade secret agreements with former employers, you should have your attorney examine those documents for possible liability.

Copyrights

Copyrights traditionally gave protection from copying literary, dramatic, musical, and artistic works. There is an evolutionary process tak-ing place in copyright law as it applies to high-technology. The development of personal computer software in the 1980s has forced the courts to take a new look at the law to determine what is protectable. There are two major issues: the protection of the basic code, and the "look and feel" of the new product. If the basic code is original work, it is clearly protectable under the current copyright law. The "look and feel" describes how the software is viewed by the user. Issues of convenience, simplicity, and graphic presentation of commands are part of the "look and feel" of software. Xerox filed a lawsuit against Apple Computer claiming that Apple infringed its copyright on display and command technology when the Macintosh and Lisa computers were developed. Later, Apple sued Microsoft, claiming that Microsoft's Windows software infringed upon the Macintosh's user interface. Lotus Development Corporation has filed a number of lawsuits against companies who have designed clones of their popular Lotus "1-2-3" spreadsheet software, claiming infringement on the "look and feel" of its product. At the time this book was being written, none of these "look and feel" issues had been decided by definitive court decisions. Until such decisions are made, you should be aware that if your new product imitates the user interface of a popular software product, you may be involved in a copyright lawsuit. Hopefully, these issues will be decided soon.

Copyrights are enforceable for 75 years from the first publication or for the life of the last author to die plus 50 years. They are enforceable only against people who you can prove copied the work; a copyright cannot be enforced if the product was developed independently.

A copyright notice is not required for protection, but the notice is advisable since there are additional rights that flow from its use. However, the copyright must be registered prior to suing for infringement. There are also some advantages to registering a copyright in the case of a lawsuit:

1.   If the copyright is registered prior to publication, within three years after the first publication, or prior to infringement; then the court may consider granting a preliminary injunction within six to nine months of filing the lawsuit. Trial will typically take three to four years after the filing of the lawsuit.

2.   Attorney's fees may be awarded in successful litigation. Maximum penalty in copyright cases is $10,000 per infringement; the minimum is $250. The penalty is per infringement, but there is a question of what constitutes an infringement. For example, if there are 100,000 copies of a piece of infringing software in use, is that one infringement, or 100,000?

The copyright symbol © should be placed on everything, such as manuals, marketing materials, software, printed circuit board artwork, and even semiconductor masks. The registration of a copyright is a business decision, but applying the notice provides some protection. Don't file the copyright if you think it will be denied, because the denial is a public record that can be discovered. Software that is copyrighted must be readable. Object code, the code that runs at the machine level, is not readable; source code, which is usually written in a higher language, is readable. Consequently, when software is copyrighted, the source code is divulged to the public, since the records can be obtained from the copyright office at the time of application. The copyright office has rules of minimum disclosure and can be petitioned to permit the filing of the registration application without a copy of the work. The rules provide for the filing of as little as the first and last 15 pages of code. The petition route is not often successful, however.

Trademarks

A trademark is a word, name, symbol, design, combination of word and design, or slogan used by a manufacturer to identify its products and distinguish them from the products of others. Trademarks can be a useful marketing tool where there is some product differentiation. Apple Computer has used the trademark very effectively to differentiate their computer for the common person from the traditional data processing image of minicomputers and mainframes. (They have also had some problems with a lawsuit from Apple Records because of their development of multimedia capability for their computers, which Apple Records feels infringes on their trademark rights in the musical area.) There are four categories that determine the level of trademark protection:

1. *Generic of Product.* There is no protection for this category of marks. A name that did have trademark protection can lose that protection if the name becomes generic of the product. For example, Bayer still has protection for "aspirin" in Canada, but it has become generic in the United States. To purchase a headache remedy other than Bayer Aspirin in Canada, you must ask for a "headache powder" or another descriptive item. Xerox is in danger of losing its mark since the mark is becoming a generic term for copy machines and photocopying, and is now mounting an aggressive campaign to protect its trademark

rights. In the past, the company has misused the mark by using it as a noun—Xeroxography—and as a verb—to "Xerox" a document. A trademark should only be used as an adjective and never as a noun or verb to maximize its length of protection.

2. *Descriptive of Product.* Marketing people like this because it implies a use of the product; lawyers don't like it because it is difficult to register and protect. If a descriptive mark is used for several years, it may take on a secondary meaning and become protectable.

3. *Suggestive of Product.* Marketing people like this less, and the lawyers like it a little more. These marks are related to the product but that relationship only becomes obvious when the product is known. For example, Safeway has a line of cleaning products that it markets under the name "White Magic."

4. *Arbitrary of Product.* These are the easiest to protect. The court will carefully look at the mark and if a competitive mark is introduced, the judge will disallow it. This type of mark can also be difficult to establish as a marketing tool. Extensive advertising and product or company promotion can be required to establish the mark's identity. Many of the marks in this category are coined words (such as Kodak) or are pure designs without words such as the "wool mark."

You can file for trademarks you intend to use. However, an affidavit must be signed that you intend to use the mark under the penalty of perjury. There is no limit on the number of marks on a product; this is strictly a marketing decision.

A technique that can be used to develop a strong arbitrary mark on a new product is to put two marks on the product, one descriptive, the other suggestive or arbitrary. Phase the first one out and phase the second one in over time as the product gains market presence.

Trademarks can be protected as long as they are used. They can take just about any form: pictures, designs (one of the best to protect), names, or type styles. Words are more in danger of becoming generic if the product becomes popular.

In summary, the judicial system that adjudicates patent, copyright, and trademark cases has changed dramatically since 1982. The law is constantly changing so you should be aware of how to protect your rights by consulting a knowledgeable patent attorney for expert assistance.

The Big Differences in OEM Businesses

At first glance, wouldn't it seem like a good idea to have only a few customers who buy your product in quantity, incorporate it into their own products, and sell to a broad base of end users? A large segment of the electronics industry does just this. Such a form of business is called an OEM business. "OEM" is the acronym for "original equipment manufacturer," referring to the company that sells to the end user. Although sales costs are lower with this approach, there are many hidden risks associated with it.

Examples of high-tech OEM businesses selling their products exclusively for incorporation into other products are single-board computers, operating system software, and disk drives. Less obvious examples are PCs, measurement instruments, and bench power supplies, which can all be sold either directly to the end user or to an OEM for incorporation into a larger system. In these latter cases, a generous discount is granted to the OEM. In return for the discount, the customer signs a purchase agreement promising to buy a defined quantity of the product and certifying that he is an OEM and not just buying in quantity for his own use or for resale. OEM discounts tend to range from 20% to 50%, reflecting the costs that OEMs incur in stocking and reselling the item incorporated into systems in lower quantity.

For most of this book, we're assuming that you plan to manufacture a product which is to be sold to the end user. The principles we outline generally apply to OEM businesses as well, but there are some major differences between these two types of business.

OEM businesses usually have a long, slow start-up pattern of their sales. Customers must first evaluate the product and then design it into their hardware, and that takes time. It often takes two to five years for evaluation sales of an IC to result in quantity orders, and other OEM products are nearly as slow. An unsettling pattern in some OEM sales curves is a bulge of evaluation units, followed by a period of lesser sales, growing to mature order levels years later. It takes courage to be in a business like that! If your product is a compatible replacement for an existing product, you may be able to move quickly from evaluation to mature sales. An example of such a replacement product is the disk drive for personal computers.

OEM sales are often highly cyclical. This phenomenon is termed the "OEM effect," and it is related to your customers' inventory. Orders for your product at the beginning of its life may be higher than expected, because some of the units are going to fill the inventory pipeline. Likewise, if your customer's business grows, additional orders will be placed to increase inventory levels. On the upside, the OEM effect is quite beneficial. But on the downside, it is devastating. When the customer's business turns down, not only are fewer units needed for production, but also fewer units in inventory. The orders placed with you can go to zero for a period of time due to a seemingly mild reduction in your customer's schedule.

If your business is an OEM business, you must build according to customers' forecasts of their needs. Accordingly, you will always be vulnerable to an "OEM effect" catastrophe if the customers' forecasts are inaccurate. This unsettling situation is *inevitable* in an OEM business.

It is common for an OEM business to be dependent on a few large customers. Although this makes your sales task simple, it creates a situation of high risk. If one or more of your customers precipitously stops buying from you, it can threaten your survival. Domain Technology, a Silicon Valley manufacturer of hard disk media, fell victim to the rapid closing of its major customer, Miniscribe. Not only did Domain's orders fall dramatically, but the company was also unable to collect its large account receivable from Miniscribe. As a result, Domain was forced out of business.

A different problem occurs when customers place duplicate orders because they see deliveries becoming dangerously long. A good materials manager insures the availability of needed parts, and many of them hedge their bets in this way. The IC industry lives with this problem constantly. . . and nearly dies from it every so often. In other industries, cancellation charges are used to preclude overbuying, but most electronics companies have not been able to make cancellation charges stick.

It is common for OEMs to establish second sources for their most critical parts. From the viewpoint of the purchaser, second sourcing makes sense to insure a steady supply of needed parts. So, if your business is to be OEM-oriented, you should prepare a second source strategy. Successful strategies include partnering with a similar company, exchanging marketing and manufacturing rights with a foreign company, licensing your designs to other companies, and granting the customer access to the design for limited internal manufacturing. It's important to determine whether your business is perceived by customers to require second sourcing and to act on that information early in your company's history. If you don't, your customers may do it for you, but in ways you won't like. For example, they might buy half of their requirements from your direct competitor or encourage another company to enter the business.

There is a also hidden competitor in the OEM business—your customer's research and development department. Usually, your customer has decided to buy from you as part of a careful make/buy decision. The decision to "buy" rather than "make" is reversible, however, and rarely will the supplier be consulted when the decision is revisited.

Sales costs are likely to be lower than selling to end users, since you are selling quantities to the same customers year after year. At the same time, prices are lower due to the OEM discount. The result is often an apparently high percentage of cost of sales. These financial characteristics force the OEM business into a low-cost mentality, where low expenses are necessary and reductions in manufacturing costs are a way of life.

The marketing department of an OEM business has a special challenge. The end user who creates the demand for your product probably doesn't know the name of your company, since your product is usually hidden. As we discussed earlier, recognition of your company's name is crucial to its success, so your marketing department must develop innovative means to promote your name. This was a challenge faced by Intel, manufacturer of the microprocessors used in virtually all MS-DOS personal computers. Intel's response was their "Intel Inside" campaign which promoted the fact that an Intel processor was the computational heart of most PCs. Intel even went so far as to run "Intel Inside" television commercials during major sporting events on network television during 1992.

Dealing with the Differences: Five Rules to Live by

Having identified all these differences in the character of OEM businesses, let's discuss some ways in which the start-up company can work to accommodate them.

1. *Know the end user.* Even though you do not do business directly with the end user, it is important to understand his business. This will help you develop a better product strategy, for you will be armed with information about your customer's market when it comes time to discuss new designs. Knowledge of the end user market can also help in understanding, negotiating, and using your customer's sales forecasts. Quanta-Ray landed a nice contract to supply lasers to a medical equipment company. The lasers were to be used in a system for eye surgery. Rather than allow the customer to dictate the characteristics

of its marketplace, the marketing team at Quanta-Ray set out to become experts in the market for eye surgery equipment. They attended the major trade shows and read the appropriate trade journals. When the marketplace became overcrowded with new eye surgery systems, Quanta-Ray was able to predict that the system manufacturer's forecasts were too high, and set a conservative production schedule.

2. *Stay in touch with developments in the customer's industry.* The more you know about your customers, the better. By staying close to them, you may be able to avoid business surprises like the customer-created second source or the internal design which could displace your product. It is also important to understand technical developments within your customers' company and throughout the industry. This information is invaluable for developing new products for the industry, perhaps using their own technology.

3. *Develop a close business relationship with your customers.* There is a clear trend in high-tech industry toward closer relationships between vendors and customers. Sole source purchase agreements, long-term contracts, and cooperative development agreements became commonplace in the 1980s and continued through the early 1990s. Behind this trend is an understanding that true quality arises from cooperation, not confrontation, with vendors. As a participant in an OEM business relationship, you should strive for the greatest degree of cooperation with your customers. You will benefit in many ways, not the least of which is an accurate sales forecast.

4. *Strive for the highest standard of product quality.* Yes, quality is important in any manufacturing business, but it is especially crucial in an OEM relationship. Your product may be only a small element of your customer's system, yet a failure can shut down the entire system. So quality must receive special attention on your part. This news isn't necessarily bad. Product quality is an exceptionally strong source of competitive advantage.

5. *Plan your business conservatively.* We have identified a number of unique risks in OEM businesses, most having to do with unpredictable order patterns. Be conservative in projecting future business levels. Remember, you can never be sure how much the customer will order, and the downside leverage in an OEM business is large. The good news is that there is also leverage on

the upside with a conservative business plan. Done correctly, the plan will yield extraordinary profits in the good times. Save them away for the bad times.

Help! Getting Outside Assistance

Starting a business is a new experience for most entrepreneurs. Many questions must be answered, such as defining the product, evaluating the market, determining financing requirements, and finally writing a business plan. But don't get discouraged—there is help available to get you started, and often it's free.

Professional services from attorneys, accountants, and consultants are available to help develop your plan. Sometimes they will provide advice as an investment in a new client; however, consultants usually expect to get paid for their services. Your banker can also help in the planning stage since he or she is interested in developing new customers.

You should carefully evaluate how manufacturing, marketing, and product development will be done in your company. Will you manufacture most of your products or make extensive use of contract manufacturers? Will all product development be done by your staff or can outside consultants meet your needs? A vast infrastructure of contract services exists across the country and throughout the world that can provide sophisticated capability beyond your size. For example, custom integrated circuits can be designed by consultants and manufactured in a world-class fabrication facility in the Far East. Focus on those capabilities that will give you a better competitive edge rather than duplicating capability that is easily purchased.

Once your plan is written, the challenge is to raise the financing (we will cover this in our next chapter). Your network of friends and acquaintances can be invaluable in making introductions to private investors or venture capitalists. You should certainly involve your attorney in the process to help you make the best deal and to avoid potential legal problems with private placements.

Outside assistance can be more critical once the business gets started because there usually aren't enough people with the required expertise in the company. Besides legal and accounting services, there are many other areas where a job may be better handled by outside services than by company employees. Some control is relinquished, but for jobs that require special expertise for a short time, or extensive capital investment, a reliable vendor may be the best choice for the job. Some companies are finding that even product development can be done effectively with outside consultants—you eliminate those periods when engineers are between projects.

There are a variety of marketing services available to help your company get its message to its customers. There are full service agencies that can develop advertising campaigns, design brochures, create public relations programs, and do market surveys. There are also specialty houses that provide one or more of these services. Even the largest companies use outside marketing services extensively to maintain freshness in their promotions and to minimize fixed costs.

Your outside resources can also help to mitigate the stress that often accompanies the start-up experience. They can provide valuable insight into problems or opportunities that otherwise might bog you down because of overload or myopia—or both. The board of directors should be used to help resolve strategic and major operational issues. Many CEOs like to have an informal advisory board review their ideas before they present them at a formal board meeting. This need can be satisfied by your network of friends and business associates, but there are also organizations, such as The Executive Committee or The President's Club, that exist for the sole purpose of helping CEOs keep things straight. Check the telephone directory to see if a chapter is located near you.

Consultants are another resource that should not be overlooked when a special need arises. Choose them carefully and look for a track record and chemistry that works for you. We recommend that you focus on candidates that have had actual business experience and worked with businesses of your size. Much of what you will be looking for cannot be learned in the classroom.

One of the most important outside relationships you will develop is with your banker. He or she should be treated like another member of your team and be kept informed of all major decisions that could have any material effect on your company's performance. Confide in your banker on new ventures and current activities in the company.

Starting a business is a challenging and difficult process. Don't get shortchanged by trying to do it all yourself.

CHAPTER 4

Funding Your Venture

Y ou may think that you're starting a company to build a product, but much of the time you're going to think that you're in the financing business! Other than recruiting quality people, securing money to fuel the growth of your company will probably consume the major portion of your time. Money is needed to finance product development, inventory, payables, and receivables. It must also be available when you need it. The good news is that there are a lot of alternative ways of raising money.

Let's start by looking at the requirements in the start-up phase. The financing you are able to obtain will be related to the future size of the company. If your vision is a $5,000,000 to $10,000,000 company in the next five years, venture capital is clearly out of the picture. Most venture capitalists are only interested in companies that can grow to $50,000,000 to $100,000,000 within five years in a market that may be several times that size. You can't expect to raise $1,000,000 for a $5,000,000 company, so size the investment you are seeking to the expected scale of your company.

Do you want management help from your investors? This is often an important ingredient in seeking investment and is not usually provided by nonprofessional investors such as friends and relatives. Consider this carefully when evaluating investment money, particularly when dealing with professional investors such as venture capitalists.

How will your investors get repaid? This is often overlooked by many entrepreneurs seeking start-up capital. Your deal only looks good to an investor if the return and payback scenario are attractive.

Raising money is an ongoing activity and your business plan should reflect the times at which additional funds will be needed. There is a truism in the business world—*don't raise money when you need it because then it's too late.* It also takes more than planned; a good rule of thumb is to raise 50% more than you think is necessary to take care of contingencies. If your plan calls for second and third rounds of financ-

57

ing, be sure that your initial investors understand that additional funds will be raised in subsequent rounds so that future dilutions of stock are not a surprise.

Try to get the highest possible stock valuation on your company to minimize dilution of the founder's stock and to leave room for future rounds of financing. There are many cases where company founders retained an insignificant portion of the company due to dilution from additional financing. Additional funds should be raised at higher stock valuations to keep the initial investors happy, which usually requires that the business be growing in a tangible fashion. Stock valuations are very fickle; new trend businesses such as personal computers and associated products achieved very high valuations in the early 1980s, so excitement about your product and industry plays a big role in valuation.

The financial climate also has a significant bearing on stock valuations and your ability to raise capital. If U.S. Treasury interest rates are at 15% as they were in the early 1980s, investors will be looking for a much higher return on risk investments than if Treasury rates are at 5%. Capital availability is also subject to the general financial climate. For instance, as the 1980s came to a close, the U.S. banking system was straining to stay profitable and federal regulators were scrutinizing bank activities very closely. This situation limited banks' lending activities, which allowed many venture capitalists to assume an investment banking role by funding later stage businesses. As a result, the amount of money available for start-ups was significantly reduced.

The business plan is your selling tool for raising money. It should describe the financial expectations for the company in detail in a form that is according to generally accepted accounting principles (GAAP). Most investors recognize that the financial projections are not cast in concrete, but they should reflect the most probable scenario you can imagine.

The dominant focus of most business plans is on the market and product, but experienced investors place their bets on people more than product ideas. History has repeatedly shown that good people can take a mediocre product and make a successful business; incapable people can take even the best of ideas down the tube. So if you think you have a great new idea and are having trouble raising the money to finance it, take a close look at the founding team. You may have to make some changes.

Be sure that you understand the capital structure that you want to establish for your company. How much equity and how much debt financing are you going to need? What classes of stock are you going to issue? You will usually want more than one class of stock so that insiders can receive stock options at a lower price than investors paid. This can be handled with preferred stocks of different classes and a common class. Get help from your attorney and accountant to determine the best structure. If it is not done right, you may have trouble raising money in later rounds of financing.

But let's now examine the Big Question: *how do you get the dough?* There are many alternatives for raising capital for a new business. The field is one of constant change, so what was true yesterday may not be the case today. Figure 4-1 compares the popularity of various financing sources in the 1980s with their projected popularity in the 1990s. As you can see, the potential sources of capital change over time and what has been a good source in the past may not be so today (and vice-versa!). We will discuss these sources and others in the order of probability of obtaining funding from them. Be prepared to spend a lot of time and energy in this process. If one source doesn't work out, try another and don't get locked into just one avenue for financing. You'll get your best deal if potential investors perceive competition for your stock.

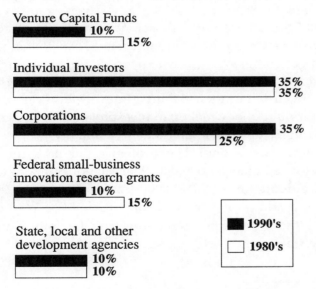

Figure 4-1: Sources of funding for a start-up company change over time, and entrepreneurs must be flexible in their quest for money.

Before we discuss the various funding sources, a short story illustrates that a "money chase" can lead to funding from the most unlikely sources. Kalok, a Silicon Valley manufacturer of 3.5-inch disk drives, was trying to raise $1,000,000 to fund the start-up of the company in 1987. Steve Kaczeus and Wayne Lockhart, the co-founders, had received $150,000 in bridge loan financing from a venture capital firm with promises of equity financing. Negotiations dragged on, but no equity funding was forthcoming from the venture capitalists. The two founders decided the time had come to try other avenues, and cut the tie with the venture capital firm. Now they had no financing, a bridge loan to repay, and only a small amount of money to keep the project going!

In the past, Steve had hired a Japanese mechanical engineer who was working for Seagate in Japan when Steve launched Kalok. Steve confided to that engineer that he was trying to start a new disk drive company but was having great difficulty raising money. The engineer suggested that they contact his former professor, a Japanese national, who was interested in investing in an American company but had been continually upstaged by other Japanese investors. The professor was contacted and arranged to meet with the two entrepreneurs in their borrowed office space.

On the appointed day, the professor arrived at the office and met with Steve to review the design of the drive. The professor poured over the drawings for two hours. After reviewing them he raised his head, turned to Steve and said "This is very good. I'll loan you $2,000,000." Steve was beside himself and couldn't wait to tell Wayne that the professor was going to lend them $1,000,000 more than they needed to get going. When Wayne heard Steve's story, he said "Steve, there's a few things I have to explain to you. First, Japanese professors are highly regarded in their country, but they don't have this kind of money. Second, nobody makes an investment of this magnitude without a thorough study of the market and company."

But Wayne was wrong! The professor did have the money and eventually loaned them $1,000,000. The relationship with the professor opened the doors for a $20,000,000 financing package for Kalok. As you pursue funding for your company, remember Wayne and Steve. Sometimes there are true angels that can be found in any of the places that we discuss.

Personal Investment

This is the source over which you have the most control—your own financial resources. Personal investments by the founders are usually part of the initial seed money for a new business. Even if you can obtain other financing, most investors like to see the founders have some of their financial hide on the line. Putting up early money to get a prototype built to prove feasibility of a new product is a good way to keep valuations higher when outside financing is sought, and may be necessary to get any financing at all.

Friends and Relatives

If you have a rich uncle, give him a call and tell him that you have the financial opportunity of a lifetime. Friends and relatives are often willing to put in seed money to help get a business going. But if you need additional capital, they may not be inclined to increase their investment. You should be prepared to tap other sources if additional funds will be required past the start-up phase.

Private Individuals

According to the *Wall Street Journal* in 1989, private individuals or "angels" backed more than 30,000 new businesses per year compared to only 2,000 for venture capital firms. Of course, many of the businesses backed by private capital are much smaller than most of the venture backed businesses. There are organizations such as the Venture Capital Network at the Massachusetts Institute of Technology and Innovative Capital Partners (Waltham, MA) that raise private capital for new businesses. There are also consultants that will introduce you to wealthy private investors. You will have to pay a finder's fee for money raised through such organizations and consultants, which can range from 5% to 10% of the amount raised. Private investors will usually not provide any management or business expertise and may not have deep enough pockets if you need more than one round of financing.

Venture Capital

Venture capital is treated in depth later in this chapter. Venture capital is one of your best alternatives if you have an experienced management team and an idea that can grow to a $50,000,000 company in five years. It also helps if you have done virtually the same business for your former employer.

Customer Financing

This is an area many potential entrepreneurs overlook. If your product is unique and unavailable to your potential customers in other forms, they may fund the start-up of your business. Megatest, a Silicon Valley manufacturer of LSI logic test equipment, funded the start-up of their company with a development contract from Intel. Another entrepreneur that worked for a communications company funded the start-up of his company with the help of his former employer's customer when they dropped a communications product line that the customer needed. This financing can have very favorable terms and may even provide an exit scenario through a sale to the investing company.

Government Contracts

The federal government has helped fund many new companies through development contracts for products. This can be an attractive way of getting start-up money, particularly if you have a technology that can be applied to a contract for a product and if you're willing to plow the profits from development contracts back into the business.

There are funds available from all federal agencies through the Small Business Innovation Research Act (SBIR). This mandates that each agency set aside a certain amount of their research and development funds for SBIR contracts. SBIR grants typically start at $50,000 for a study contract, but can be expanded in follow on phases to over $250,000 to bring a product to market. Contact the agency that may have an interest in your technology to get their catalog of SBIR opportunities.

You can also get development funds by making an unsolicited proposal to an agency or responding to a published request for proposal. Federal requirements are published in the "Commerce Business Daily" distributed by the federal government to subscribers. It helps to know the contracting officer so that your proposal will get a fair hearing.

Corporate Financing

Many large companies have started their own venture capital funds to finance start-ups that can innovate faster than their own culture allows. There is financing available in the United States as well as extensive sources abroad, particularly in the Far East. Many Japanese companies have made early investments in high-technology start-ups to gain access to new technology. For example, Kubota made a large investment in Stardent, a supercomputer start-up that could not raise funding in the United States. This can be a particularly attractive source of funds because the investor can lend business advice, open doors for strategic alliances, and provide deep pockets.

Government Enterprise Funds

Many states and localities have established enterprise funds and facilities (such as the Research Triangle Park in North Carolina) to attract new businesses. In Canada, the province of Alberta has established a fund that actively invests in new businesses like a venture capitalist.

The federal government is also active in this area. According to the San Jose *Mercury-News*, the Defense Advanced Research Projects Agency (DARPA) made a $4,000,000 investment in Gazelle Microcircuits, a manufacturer of gallium arsenide semiconductors in 1991. New legislation passed in 1989 allows DARPA to not only award development contracts, but also to act like a venture capitalist for technologies it considers to be essential to national security. Check out the activity in your area; there might be a funding source for your company.

Small Business Administration

The Small Business Administration (SBA) provides funding guarantees when a business is turned down by a commercial bank. The SBA offers two basic types of loans: guaranty loans and direct loans. Guaranty loans are made by private lenders, usually banks, and can range from $150,000 to $750,000. The process is initiated with the private lender who conducts the initial review and submits the application to the SBA. If the loan guaranty is approved, the private lender processes the loan and disburses funds.

An SBA direct loan has a maximum amount of $150,000 and is only available to applicants unable to secure a guaranteed loan. Direct loan funds are very limited and often only available to businesses located in high unemployment areas or owned by disadvantaged people.

The same information is required for an SBA loan as any other financing and should be contained in your business plan. The agency will back start-up situations that meet certain financial and business criteria. Consult the SBA office in your area.

Research and Development Partnerships

Research and development partnerships have been used in the past to fund development programs for start-up companies. Due to changes in the tax laws in 1986, this avenue is not as attractive as it was in the late 1970s and early 1980s. Your attorney can advise you on this source.

Universities

A university may seem like an unlikely place to look for start-up funding, but there are many opportunities to take advantage of university research to launch a new business. Many schools are actively seeking entrepreneurs to take research out of the laboratory to the commercial marketplace. For example, Ariano Technologies, a Boston start-up company, licensed fiber optic chemical sensor technology from Tufts University to launch a new firm in environmental monitoring in early 1991. This approach can significantly reduce the amount of seed money required to develop a new technology.

Banks

Banks do not lend to start-up companies unless you want to provide 100% collateral for a loan such as your house or car. This has been done many times. The typical banker requires profitability and three years in business before he will discuss lending agreements that are not totally collateralized by the personal assets of the founders. Even after your company has reached profitability, the founders may still have to personally guarantee bank loans. However, there are some unique banks, such as Silicon Valley Bank (SVB) in Palo Alto, California, that are more responsive to the needs of high-technology entrepreneurs. SVB will lend to start-up companies if they are not profitable but have a good balance sheet. The typical company is venture capital backed, well capitalized, has a product developed, and has started production. This type of financing is for working capital for inventory and accounts receivable; it is one step beyond start-up financing.

As part of your financing strategy you will want to establish a banking relationship to provide a line of credit to finance the company's working capital needs (inventory, accounts receivable, and accounts payable) when production commences. In addition to providing funds, the bank will provide credit reports to potential customers and suppliers who may be concerned about your financial strength, financial backup for large orders, equipment financing, and letters of credit for international sales. The bank will require your company to meet certain financial covenants such as profit level, working capital amounts, current ratios, net worth, quick ratio, and debt to tangible net worth. Don't be put off by these requirements; negotiate the best deal and consider the requirement to be a spur for you to keep an eye on the financial performance of your company.

Shop for your banking services like any other item. Get references of other companies that use the bank and find out how the bank operates. (Do they stick by their customers, or cut them off when the going gets tough?) Once you have selected a banker, treat him or her as another member of your team. Keep your banker informed of the good and bad things that are happening to your company. Avoid surprises at all costs, particularly negative ones.

As you arrange financing, be sure to keep the capital structure of your company in balance. The capitalization of your company has two principal components: equity and debt. Debt is particularly attractive from the stockholder's point of view because it does not dilute ownership. The downside is that the interest payments on the debt can become burdensome and force the company into a cash flow crunch.

If you need additional rounds of financing, plan for them early and consider convertible debentures that can be paid off at the company's option before conversion. This kind of security is often more easily sold to corporations than stock and requires less legal work. A corporation that owns some of your paper may be interested in acquisition at a later date. The best solution is to arrange for an adequate source of funding up front. Let's look at how this might be accomplished using venture capital.

The Upside and Downside of Venture Capital

The Good News is that venture capital is an excellent form of financing for start-up companies, for reasons we'll explain in this chapter. The Bad News is that venture capitalists can be difficult to work with, and that access to venture capital became quite difficult for start-up companies in the late 1980s and will probably stay that way through much of the 1990s. By 1990, venture capital firms had established a clear trend toward fewer and larger investments, made mostly in later stages of companies' growth. The results (at the time this book was written) of this trend are a lack of seed money for new companies and greater competition among entrepreneurs for later stage funding.

Some people consider venture capitalists to be the "bad guys" of high-technology industries. They grill people mercilessly, they don't return phone calls or business plans, and they are often observed reorganizing or closing companies down. This reputation is overstated. Venture capitalists are investors and make important financial—not personal—decisions. The amounts of money they work with are large, so they need to be extremely careful about their decisions.

They are besieged by mail and telephone. "Money is like flypaper," one has told the authors.

In order to deal with venture capitalists effectively, you need to understand them. Unlike most funding sources, venture capitalists will get involved in their investments. In fact, they spend more time with their portfolio companies than with new investments, since the performance of their investments is the key to their financial success. The typical venture capitalist sees five business plans a day, yet only invests in two or three each year. Having made numerous investments, he or she is likely to serve on several boards of directors. The pressures of the job are enormous.

Venture capitalists rarely make solo investments, but prefer to team with other funding sources—usually other venture capitalists. This posture is merely a form of risk management. Also, groups of venture capitalists often find themselves chasing the same technological innovations, as several different start-ups vie to be first in a new market. Given this behavior, one venture firm will become the "lead" investor in any start-up, performing the final negotiations with the entrepreneur and making the largest investment. Once you begin discussions with venture capitalists, it is important to push for a lead investor commitment from a larger firm. Resolving this issue among several venture capitalists can take a long time.

Venture capital firms prefer to make their investments in phases in order to minimize the risk of their capital. The most common phases are:

Seed	To $500,000
First round	$1,000,000 or more
Expansion	Several millions
Mezzanine	Even larger

A classical case of successful start-up funding is that of Mentor Graphics, the computer-aided-engineering systems manufacturer in Oregon. Mentor received seed funding of $1,000,000 from three venture capital firms in 1981, first round financing of $2,000,000 one year later, and expansion financing of $7,000,000 in 1983. The company went public in 1984, raising $55,000,000.

Venture capital firms tend to specialize in some way. Some are industry specialists—for example, medical equipment and services. Some specialize in taking the lead at certain stages of investment. Many firms concentrate their investments in their local geographic area, or in several areas if they have offices in multiple locations. It is wise to learn these characteristics of venture capital firms early in your exploration of this funding source.

Venture capital firms take an intermediate-range view of their investments. The partnerships by which they raise funds have a life of about ten years, so a venture capitalist must select investments, help the companies to grow, and sell out or go public within a decade. Therefore, venture capitalists are biased toward business plans which show the start-up being acquired or going public around the fifth year, which allows sufficient time between the transaction and selling the stock to satisfy legal constraints and await favorable market conditions.

As mentioned earlier, venture capitalists generally only want to see big plans, like a projected company of $50,000,000 annual sales selling into a market several times that size. They generally cannot afford to get involved with smaller operations. The return a venture capitalist seeks is about ten times the amount invested within five or six years from the investment. This return may seem outlandish, but the high failure rate of start-up companies causes it to average out about right. In the heyday of venture capital—the early 1980s—only 5% of investments actually made returns of ten times or greater.

In your business plan, the venture capitalist will look for three important points beyond finances: management team, strategy, and technological edge. Of these three, the management team is by far the most important, so be prepared to undergo considerable grilling from venture capitalists on the qualifications of your management team. They will be looking for related management experience and perhaps "training" inside big companies. You should check your team's references yourself to insure only positive responses. You should also insure that the CEO identified in your plan is of suf-ficient stature to satisfy the venture capitalists' tough requirements. This news is not all bad, since venture capitalists are an excellent source of contacts to help you fill key positions on your team.

Advantages of Venture Capital Funding

▼ *Venture capital firms have access to large sums of money.* It is best to have initial investors who are capable of larger investments in future rounds, so venture capital firms are ideal in this regard. If a venture capitalist cannot commit additional funds later on, he or she can often find other venture firms to assist.

▼ *Venture capital firms offer prestige to their portfolio companies.* This is especially true of the larger, well-known firms. Prestige can mean a lot to a start-up company when it comes to borrowing money, entering into a lease, or incurring other financial obligations. It can also have a beneficial effect in relationships with customers, vendors, or other business partners. Certainly, prestige will count heavily when negotiating an IPO, acquisition, or other form of financial exit.

▼ *Venture capitalists can usually offer management assistance to the start-up company.* This is a strong argument in favor of venture capital. We caution you, however, to check on the reputation of each firm in this regard. The track record of venture capitalists rendering assistance to their portfolio companies has been spotty.

▼ *Recruiting is an area where venture capitalists can be of great help.* They have not only a large number of personal contacts, but also a file of resumes.

▼ *With their broad business contacts, venture capitalists are often good sources of alliances with other companies—vendors, distributors, OEMs, etc.* You could find this help invaluable if your contacts are limited. Venture capitalists sometimes even have contacts among potential customers.

▼ *Venture capitalists are helpful when constructing your financial exit.* They have excellent contacts in the financial communities and they've done it before. (See "Planning for Liquidity with an Exit Scenario" later in this chapter.)

▼ *The most valuable help venture capitalists supply is advice.* They are expert on business strategy since they live with it every day. They are usually quite knowledgeable about business finance. Since many venture capitalists are former business executives themselves, they can bring experience to bear on your operational problems.

Disadvantages of Venture Capital Funding

▼ *Because they are so busy, venture capitalists don't always have the time to give the start-up company the level of help it could use.* Managers of start-up companies sometimes find themselves fighting for the attention of their investors. This can be avoided by investigating the reputation of venture capitalists before making funding commitments. Also, be sure to check that a venture capitalist who might be elected to your board of directors isn't already on a large number of boards.

▼ *Quite the opposite to the point above, venture capitalists have a well-deserved reputation for pushing their weight around.* Such heavy-handed behavior as surprise firings of company officers or last-minute decisions on funding happens all too often. You should ensure that venture capitalists with whom you're dealing are people you feel comfortable with.

▼ *There is no escape from the demanding due diligence process to which you will be subjected by the venture capital community.* You will be asked the toughest questions repeatedly. The process often drags on for months, as the venture capitalists involved each verify your references and check your story using their own sources. All this is designed to uncover every detail of your plan and your people, to insure that the venture capitalists have a complete understanding of the risks involved. Basically, you have no choice but to cooperate fully. The more open you are, the sooner the process will be finished.

▼ *Venture capital firms typically demand a larger share of equity in your company than other funding sources.* In a first-round investment, venture capitalists will require 30% to 60% ownership of the company. The rationale for this posture is that they offer more value to the start-up company than other, less involved, investors. For the most part, this rationale is true. The key parameter to keep in mind is the level of equity you will own after the final round of investment. A typical result for a successful start-up is 20% ultimate ownership by the founders and employees. Since venture capitalists often contribute the additional value they claim, their requirements for greater equity should be honored. This is not to say that you shouldn't negotiate with them—of course, get your best deal. But recognize two things. First, the equity granted an investor is proportional not only to the funds invested, but also to other value contributed. Second, all investors want the founders and employees to own a significant share of the enterprise so as to keep them motivated. Your investors know that you are the key to the success of their investment, so they're not going to allow you to be demotivated.

How to Approach the Venture Capital Community

Imagine an auditorium filled with venture capitalists and other start-up investors. You are given a few minutes to stand before them and explain your business idea. Then, later in the day, you hold court in a conference room as those venture capitalists who were interested in your presentation drop by to talk. This scene isn't a fantasy. It happens several times each year at various locations around the United States. Probably the most famous of these conferences is held each January in Monterey, California. (For information on participation, contact the American Electronics Association in Santa Clara, California.)

Something like the Monterey conference sounds like an ideal start on your road to funding. But don't expect to obtain funding commitments from a half-dozen venture capitalists there or at a similar conference. For most entrepreneurs, it is only one step in a long process. Let's back up a bit and start at the beginning.

The first step in approaching venture capital sources is to do your homework. Your objective is to research the 250 venture capital firms and equal number of similar funding sources in the U.S. to find about 50 who are a potential fit with your needs. Look for firms that specialize in your industry or technology. Look also for firms who tend to make investments of the size you seek, and at your company's financial stage. Try to find out which of the firms you learn about are possible lead investors. Then learn what you can about the experience, the portfolios, and the degree of success of the various firms.

Names of firms and their principals are available from the National Venture Capital Association and, for those on the West Coast, the Western Association of Venture Capitalists. Another good source is *Pratt's Guide to Venture Capital Sources*, published by Venture Economics of Wellesley Hills, MA. This is the most comprehensive source of information available on the industry. A wealth of information is also to be found in *Venture* magazine. Talking with experienced entrepreneurs is very helpful. Finally, you should exercise your personal acquaintance network vigorously to learn what you can from them.

It is fruitless to mail business plans cold to venture capitalists. Your odds of gaining attention are increased an order of magnitude with an introduction. So, you need to find people who can introduce you to five or more venture capitalists on your list. Finding these sources may require digging several layers beyond your immediate network, using referrals upon referrals. That's the way it is in the world of high finance. (Of course, attending something like the Monterey conference can get you a big head start.)

Once you have started to contact venture capitalists, it is possible to have venture capitalists themselves provide the introduction to the next firm. The objective at this point is to continue to make contact with venture capitalists until you are in serious discussions with three or four.

Following the introduction, your first contact with a venture capitalist should be by telephone. The objective of the call is to set up an appointment. Once the meeting is set, ask the venture capitalist if he wants you to send your business plan ahead or bring it to the meeting.

From this point on, you must be quite flexible with your schedule. Your availability for appointments is a clear indicator of your devotion to the enterprise.

At the first meeting, the subject should be you and your business idea, not funding. There will be plenty of time to discuss money later. First, you have to demonstrate the worthiness of your idea and the enthusiasm of your team. Expect the venture capitalist to ask lots of questions about subjects you're not really prepared to discuss; the firm will be digging hard for data in the first meeting. Your objective in this first crucial meeting is to establish the basics of a sound business relationship. Even so, don't be put off if this first meeting is with a junior member of the firm.

At subsequent meetings you can begin to pursue your agenda as well as deal with the venture capitalist's extensive agenda. Make sure that there is always time for a two-way conversation, for you want to ask questions like: Have you invested in similar companies? What could your firm do to help us, once we are started? Would you be comfortable as the lead investor in my company? What amount are you prepared to commit?

In these meetings with venture capitalists, you will be required to demonstrate thorough knowledge of all aspects of your proposed business. Favorite subjects of venture capitalists are market data, patents, competition, and technical abilities of your product development people.

A venture capitalist will often ask for some form of customer testimony on the utility of your product idea. This is when your market research pays off. Of course, it is preferable for a customer to have used your product, but venture capitalists do not require you to have constructed a prototype. They recognize that you need to secure funding before undertaking expensive product development.

The prototype question deserves consideration on your part, however. If there is no other way to validate your idea, then you may need to build hardware before raising much capital. In this case, you should either self-fund your product development, seek private funding for it, or possibly get modest seed money from specialized venture capitalists. Initial self-funding has the advantage that venture capitalists will allow you greater ownership of the company after their investment, since your company has greater initial value.

At best, obtaining venture capital funding is an arduous process. Be prepared for rejection along the way. Learn from each experience, and don't take rejection personally. Even when things look like they're headed for failure, don't let your discouragement—or, worse yet, panic—show. Be sure to discuss developments with other members of your team.

We recommend that your team include an experienced financial executive to help you through the negotiations. This "part-time chief financial officer (CFO)" can also help with the preparation of your business plan and other matters. It is essential that he or she have experience in start-up funding situations. Your attorney is a second source of valuable assistance.

Maintain Good Investor Relations

Closing funding with a venture capitalist consortium or other source is just the beginning of a long relationship with your investors. You now have the responsibility to keep them informed and happy with their investment.

Maintaining good relations with investors is not difficult. Two key elements are a thorough annual shareholders' meeting and a report from the president about once per quarter. Your board of directors should determine the frequency and type of communications; after all, they represent the shareholders.

Relations with venture capitalists are a different story. These investors will interact with your company officers actively. They will require monthly financial statements. A venture capitalist-funded company will need a full-time chief financial officer to oversee investor relations along with supervising the finance and accounting functions. The CFO can expect to devote much of his time to investor relations. As with all investors, the key to good relations with venture capitalists is no surprises. Always inform your board whenever things turn sour.

Partnerships and Joint Ventures

There was a time when electronics was such a high-growth industry that new products were almost guaranteed a position in the global market. They were voraciously devoured by eager consumers in both domestic and foreign markets. It was not difficult to obtain representation. Sales distributors and representatives were always on the lookout for new products to sell to their customers.

This has all changed. The spectacular growth of the electronics industries in the 1980s proliferated the market with thousands of new products, particularly in the high-growth office automation market. Companies are today faced with getting the attention of customers who are inundated with product offerings. They must quickly establish a position in both domestic and foreign markets and be able to serve the market when product sales take off. However, most start-ups are usually focused on getting their products to the domestic market, and rarely have the time or resources to address all markets. Moreover, start-ups do not have the financial horsepower or time to capitalize on their market opportunities. So what's the solution? For many start-up companies, it lies in partnerships and joint ventures with other high-tech companies.

There are many benefits to this approach. In addition to distribution, the partner may bring complementary products to the venture which may be integrated into a custom solution for the customer. Your partner may also provide technology or manufacturing know-how that you can use to gain a competitive advantage at home and abroad. You get much more than distribution because of the synergy between your product and your partner's products. In addition, the foreign sales management requirement is significantly reduced since a good partner will have a well-established sales capability.

A foreign partner can also help you localize your product. Foreign customers expect products to be tailored to suit their needs. This may require changes to the hardware, but most certainly will require changes to the documentation and user interface software to be successful.

Strategic partners can provide funding to help your company grow. The terms are often much better than with a straight equity investment because partners are getting more than a financial interest in the company—they're getting additional market penetration for their own products through the leverage of your expertise. However, investment in your company is usually not the first step in developing a strategic relationship with a foreign partner. There have been so many investment failures in poorly managed start-ups that many foreign companies are wary of kicking the relationship off with an equity investment. That will come after a period of successful cooperation. Besides, equity investment is not the only form of funding available. Distribution and technology licenses are other ways in which to get an early cash infusion to your company.

Strategic partners need not be limited to foreign markets. It may be equally advantageous to develop an alliance with a domestic partner to leverage complementary technologies or manufacturing. Cellular Data Inc. (CDI) is a 1988 Silicon Valley start-up whose success is dependent on both domestic and international strategic partnerships. CDI developed a technique for inserting packet data channels between the voice channel capability of the installed analog cellular radio network system. Their unique technology provides significant data handling capability for applications such as alarm sensing, mobile point of sales, environmental monitoring, and other low data rate applications without requiring more radio spectrum. Their challenge was to forge partnerships with the cellular operators that control the licenses for the service areas throughout the world. Without them, CDI had no business!

A joint venture is a variation on the partnership theme. In this case, a new company is formed, most commonly in foreign countries. Your company and the foreign partner own the company in percentages that are determined through the negotiating process. This is a more complicated solution than a licensing agreement, will take a longer time to set up, and is usually done after a company has had considerable success. There are some very successful joint ventures such as Fuji-Xerox, a joint venture between Fuji film and Xerox.

Be sure to establish your objectives before you seek out a partner. Understand what you bring to the party and what you expect your partner to provide. A clear understanding of each partner's objectives will greatly facilitate the process. Try to understand your prospective partner's business, particularly their market. This will help you make a deal with the right partner.

Making the right contacts is key to developing a partnership in a reasonable amount of time. You should avoid trying to develop a strategic relationship with a company by dealing with a middle manager. Start at the top. The deal will move down to the responsible operating manager, but if he or she knows that a top executive is very interested in the project, you will get a fairer hearing.

You need the right contacts to start at the top. You or members of your staff should have access to one of the top managers in any company that you are interested in striking a deal with or you should be introduced by someone who has access and the trust of the company. Starting with the right approach cannot be overemphasized. If you don't, you run the risk of your deal becoming shopworn as it wends its way through an organization. Another risk, particularly in technology licensing deals, is the "not invented here" (NIH) syndrome. No research and development manager wants to admit that he or she hasn't thought about a technology, and the most common response will be that the idea has already been investigated.

If you don't have access to a company that you want to approach, you can hire a consultant to provide the entry. There are a number of individuals and companies that have good, long-term relationships with companies throughout the world that provide this service. Check out their credentials carefully to be sure that they can deliver the goods. Look at their track record and request a detailed description of their relationship with the targeted company and their proposal to introduce you.

The Appraisal

Who establishes the price of a classic painting at an art auction? No single individual does. Rather, the price is established through an active exchange of opinions between an auctioneer and investors. Each of the parties have their own appraisal of the painting's value, and the market value is determined by the auction.

As an entrepreneur, you will find yourself conducting auctions to determine your company's valuation on several occasions. You must be able to perform your own appraisal to do this well.

The valuation of a corporation is its number of outstanding shares of stock and stock options multiplied by the price per share. A new valuation is established at each funding stage of a start-up company, since investors are purchasing shares at a known price and the resulting total number of shares is known.

When a company's stock is traded publicly, its valuation is set daily by trading in the stock market. In this setting, investors evaluate stock value primarily by calculating the price-earnings (P-E) ratio. This is the ratio of the stock's price to after-tax profit, or earnings, per share. In the late 1980s, mature high-tech companies' P-E ratios averaged just under 15. Companies in attractive new markets or growing rapidly could command twice that ratio.

While after-tax profit is used to evaluate publicly traded stock, pretax profit is usually used for start-up companies since their tax situations vary so widely. The average pretax P-E ratio for those same high-tech companies would have been 8 to 10, since U.S. corporate tax rates run around 35%. Other ratios investors use to calculate valuation include simple multiples of annual sales or cash flow.

The valuation of your company set in the first round of financing is important, for it will determine how much of the company you own. If investors are injecting $3,000,000 and your company is valued at $6,000,000, for example, then you and the other cofounders would own 50% and the investors the other 50%. These investors would be tacitly acknowledging that your company had a value of $3,000,000 before their investment. In succeeding rounds, both your ownership and that of early investors will be diluted by shares sold to new owners.

As a rule, investors will not settle for small ownership positions in first-round financing. They look for ownership of 20% to 60% of the company. The degree of ownership demanded by investors is generally due as much to their sophistication as to the merits of the business plan or the contributions of the founders. Private investors (angels) tend to allow high valuation and therefore low ownership for themselves. Venture capitalists generally press for low valuation and investor ownership of 30% to 60% in the first round. Other funding sources like corporations fall in between.

The valuation of start-up companies is highly subjective. Often the numbers thrown out by potential investors seem downright arbitrary. But you still need to have an appraisal method for yourself. The simplest and most common method is backward analysis. Look ahead to some point in the business plan when the company could be compared to ongoing businesses—for example, having shown a profit for over a year. Multiply pretax earnings by an appropriate P-E ratio—9 for a maturing company or more than that for a fast-growing company. The result is your estimate of the company's valuation at that future date. Then work backward using the compound growth formula with a reasonable return rate to determine today's apparent valuation.

As an example, let's consider a start-up requiring $5,000,000 in capital and assume that after six years the company has pretax profit of $5,000,000. If the company was still growing rapidly that year, you might choose a P-E ratio of 18, resulting in a valuation of $90,000,000 at the end of the sixth year. An investor looking for a return of 25% per year would calculate the present valuation of the company at $23,500,000, and a $5,000,000 investment would be 21% of that. An aggressive investor looking for a 45% per year return would calculate a present valuation of $10,000,000 and insist on 50% of the company's stock for $5,000,000.

Before you begin negotiations with funding sources, try to model the approach to valuation they are likely to use. Then, armed with your own valuation analysis, you are ready to negotiate effectively. If all this financial analysis seems oppressive to you, don't worry. It's old hat to your "part-time CFO."

The valuation established through a funding round is not the sole determinant of your ownership of the company. Bonuses, options, and some financial instruments can vary the relative ownership of founders and investors according to the company's performance. An example of a financial instrument is investors' preferred stock that is convertible to common stock via a ratio tied to the company's finances. Approach these arrangements with care, since they can severely dilute your ownership—and control—of the company if things go wrong. Your "part-time CFO" and your attorney should review proposed funding packages thoroughly and look for such risks.

You need a self-appraisal of your business not only when raising money but also when planning your financial exit. As described in the next section, an exit basically means either going public or being acquired.

An investor planning to acquire your company or to purchase stock in it will often use the P-E analysis we've described. Or, he or she might view the opportunity as a simple investment and calculate the expected rate of return. This approach is possible when your company is regularly producing profits. One form of analysis is return-on-investment (ROI), the ratio of annual after-tax profit to the purchase price. This ratio ranges from 5% to 10% for mature companies, using company valuation as the investment. For fast-growing companies, more sophisticated treatment of the time value of money is needed. Investors may then evaluate the opportunity using discounted cash flow analysis.

No matter how you construct your financial exit, you'll need a careful valuation analysis to back your negotiating position.

Planning for Liquidity with an Exit Scenario

Owning a lot of stock in a start-up company is something to look forward to. But there's a built-in problem with privately held stock—government securities regulations make it very difficult to sell.

This constraint may not seem like a problem to you, but it's a big impediment to potential investors. They want to know how their investment will eventually become liquid (convertible into cash). They also need assurance that liquidity can be achieved within a reasonable period of time, such as five years. So you have to show them a plan, termed an "exit scenario," that tells how and when this will take place. The exit scenario should be part of your business plan.

You also should be concerned about liquidity. It often happens that much of an entrepreneur's net worth is tied up in his or her company. If the need should arise to create cash for personal use, like sending the kids to college or handling an emergency, you might need to convert some of your stock into cash quickly.

There are three general forms of financial exit for the founders of a start-up company: remaining private, merging or being acquired, and an initial public offering (IPO). Let's look at each of these options from the founder's point of view.

If you choose to remain private, you can sell your stock to certain types of individuals or you can sell your stock to the company. Both of these transactions are easier said than done. Since the stock is not very liquid, it's hard to find a qualified individual interested in buying it. The company usually can't afford to buy it, although companies sometimes borrow money for this purpose, in a form of leveraged buyout. If you can find a way to do it, however, there are advantages to keeping the company privately held; it's less expensive than other options, the financial records of the company remain confidential, and you avoid the hassles involved in acquisitions and IPOs.

An interesting variation on private ownership is the employee stock ownership plan, or ESOP. In this scheme, the company buys stock back from the founders and then sells stock to the employees. It is even possible for employee retirement plans to purchase the stock under the right conditions, which is quite advantageous. This is an increasingly popular form of founders' exit for firms of all sizes.

The merger/acquisition option can be very beneficial to you and your company. In this transaction, another corporation acquires your company in exchange for cash, their own public stock, or a combination. You can usually sell the stock received more quickly than stock resulting from an IPO. Also, although you will normally need a mergers and acquisitions broker to assist you, the resulting cost is much less than an IPO.

The major advantage of the merger/acquisition option is not lower cost or faster liquidity, however; it is the synergy between your company and its new parent. Most acquisitions arise from strategic considerations like pooling of complimentary product lines or access to a superior sales organization. The value that an acquiring company sees in your firm is largely determined by the strength of this strategic synergy, and it may be considerable. Your customer base alone could be viewed as having great value. As your start-up develops, you should be constantly aware of companies that are possible acquiring candidates. Strive to establish close relations with them.

Your company can be a supplier, a purchaser of their products, or perhaps a partner in joint marketing or other arrangements.

Through an IPO, your privately held stock will be converted into publicly traded stock. The process involves retaining an investment banking firm to offer stock to the public in exchange for cash. The cash is then injected into the company. In the end, you have easily traded stock, lots of cash in the company's bank account—and a large number of new stockholders.

IPOs usually result in higher company valuation than acquisitions. Investors look at the IPO as merely a purchase of stock, and they are willing to settle for lower rates of return.

Being a public company has big advantages. There is enhanced credibility with customers and greater prestige for employees. You'll have an easier time recruiting good people, using stock options. Founders, employees, and other stockholders will have greatly increased liquidity of their investment. And you and your management team can stay in place—something which rarely occurs in the acquisition scenario.

But going public also has its downside. The public financial reports required by the SEC take considerable time and often reveal more than you would like your competitors to know. Also, the visibility of your company's financial performance and stock price can cause big problems for management during hard times—just when you don't need more hassles.

IPOs are also expensive. Even a modest IPO can cost your company $250,000 in investment bankers' fees. The process demands a lot of your time. The public scrutiny created by the required financial disclosures can be burdensome. Worse yet, the price of your offering can be heavily affected by conditions in the stock market. The public is less receptive to IPOs during periods of economic downturn, so your timing may be determined by forces beyond your control.

Which form of exit scenario is best for your company? The answer depends on several factors, first of which is the funding source of your company. A corporation that owns stock in your company might want to acquire it; presumably, the reason they invested was to complement their own business. If there is one large private investor, he might want to buy out the founders. Or, the right group of employees might want to pursue an ESOP.

The size of your company is also a factor. It can be prohibitively expensive for a company with less than five million dollars annual sales to attempt an IPO, since the work to be done by the investment banking firm is about the same as for a company ten times the size. So smaller companies usually choose the acquisition route.

A factor which you cannot predict is opportunity. Successful start-up companies are often approached by established companies about acquisition. Or, if the financial markets and your company's financial performance are right, an IPO could be an obvious move for your company some day. So your business plan can only state your preference for exit, with the understanding that circumstances might modify the plan.

Your posture toward a financial exit must be flexible. You should keep your exit plan in mind as the company grows, constantly reviewing your options. The optimum time to look for an exit is when your company's financial performance is at its best, for this is when you will receive the greatest return for the size of the company.

CHAPTER 5

Staffing Your Company

Most high-tech companies are started by two or three people that have shared a common work background and decided that it is time to strike out on their own. Your start-up team should have solid relevant experience in at least three of the following areas: marketing, manufacturing, product development, finance, and sales. The marketing and product development areas absolutely *must* be covered. The team should have mutual respect and share a similar moral and business philosophy. A lack of confidence in one of the team members or jealousy of position can undermine the company and cause bickering among the founders.

For example, one company founder in Silicon Valley had problems with the founding partners in his company. Two engineers that were critical to the success of the company started bickering over their position in the company. One of them became so obsessed with his status that he began to spend most of his time worrying about his location in the parking lot and similar status symbols rather than concentrating on the product development. This type of unrest can quickly spread throughout the company and paralyze it.

After the founding team is selected, the task of filling the management functions for the rest of your company can be addressed as the company grows. You should build a strong management team in the key functional areas to provide a good foundation upon which to build your staff. All members of the team will have to wear multiple hats in the early stages. Most start-ups do not have a full-time human resources or personnel manager, so one member of your team will have to wear this hat. By the time the company gets to 50 employees, a staff member will have to devote a significant amount of time to this area.

Senior executives that have spent most of their career in a large corporate environment sometimes have trouble adapting to the free-wheeling entrepreneurial working conditions that are typical in most start-ups. This can present a dilemma for recruiting since larger corporations are good sources for experienced personnel. Structure your recruiting interview to ferret out potential problems in this area.

Although hardworking, skillful administrators and inventors may form the core of the company staff in the start-up phase, the company cannot succeed without a strong sales function. If a star salesperson was not part of the founding team, find one to build the sales program before products are ready for shipment. This person is not an administrator—their mission is to book orders for your product through personal calls on specific customers or by motivating a sales representative or dealer group.

The types of people you hire at the beginning will establish your company's culture. It is imperative that the early key employees share a common moral and work philosophy to develop a strong culture.

Recruiting The Privates

Recruiting qualified, dedicated staff is a difficult challenge you will face throughout the development of your company. You should set a very high standard to get the best people you can attract. You can't give lip service to hiring better people. You must pay the price, such as better working conditions, more opportunity for personal growth and advancement, stock options, health benefits, and comparable pay. Prospective employees should be informed of your expectations, such as number of hours professionals are expected to work, length of work week, and contribution expected.

Most start-up companies concentrate on hiring experienced personnel until the company gets to be quite large—greater than 100 to 200 employees. After you have hired a core group of experienced people in the various functional areas, we suggest that you develop a college recruiting program to bring in trainees to the organization. Although there is an investment required to develop the inexperienced graduate, the following bonuses are realized:

▼ The new graduate will learn the way your company does business. Carry-over attitudes from previous employers that are in conflict with your standards do not have to be "untaught."

▼ New graduates will be schooled in the latest technology. This is particularly important if your company's products will be based on leading-edge technology.

▼ New graduates can bring additional enthusiasm to your organization. They can help stimulate a higher energy level with their idealism and enthusiasm to take on new challenges.

College recruiting has been the backbone of personnel development for some of America's best known high-tech companies. Hewlett-Packard adopted this policy in the 1950s and for many years recruited virtually all of their new engineering employees through an extensive college program. You don't have to be a big guy like H-P to use college recruiting. Lightwave successfully used college recruiting with only ten employees.

Executive recruiters are another alternative for hiring professional staff. The most professional types work on an assignment basis where a job specification is developed and qualified candidates are screened prior to your interview. Fee structures vary; be selective when choosing prospective recruiters. Fees can be quite high, up to 20% of first year's gross pay. Some recruiters will offer a guarantee in the form of a free search if the new employee leaves within the first year.

There are also recruiters that are basically employment agencies. They collect resumes and follow the advertisements in the help-wanted sections of the local newspaper. These operators provide little value because their prospects are not screened for your job opening. Their fees can be quite high, often as much as some professional recruiters that work on assignment. We recommend caution if dealing with this type of recruiter because their fees are generally too high for the service they provide.

The help-wanted section of the local newspaper is a familiar method to reach a wide range of people that are looking for job opportunities. However, there are a number of disadvantages with this method.

▼   It is indiscriminate in the selection of resumes that you will have to review to find qualified people.

▼   Prospects will send resumes regardless of the qualifications stated in the advertisement.

▼   There is a certain amount of natural selection that occurs that adversely affects the quality of the candidate that comes through this channel. The want ad reader can fall into the following categories: laid off because of business downturn, laid off because of poor performance, dissatisfied with current working conditions, dissatisfied with career opportunities, or seeking higher paying job. The reasons for change can often cause headaches for the new employer.

It is possible to find excellent employees through help-wanted advertisements, but be cautious when using this source.

One of the best and least expensive ways of recruiting new employees is through current employees, who often have friends that are looking for new employment opportunities. A big plus for this system is that the current employee is a fairly reliable character reference for the prospective employee. This source can be stimulated by modest financial rewards for employees that are hired through the referral and that stay for a prescribed period of time.

Employment agencies can be used to develop the clerical and production staff of the company. Agencies offer the advantage that the employee can be brought in on a temporary basis without a long-term commitment. If such employees are satisfactory, you can often hire them full time, but you are under no obligation. The cost can be higher than for regular employees, but the hiring flexibility may justify it.

Getting candidates for a position is only the first hurdle that must be cleared to fill a position. The next is the interview process, which is the most important step of the hiring procedure. An interview procedure that exposes the prospect to several people on your staff should be established to find out how the prospect will fit in with the company and measure the depth of his ability. (In its formative years, Tandem spent over fifteen hours interviewing new prospects.) The hiring manager should have the final decision, but input from an interview team can often uncover potential problems that a single interviewer might overlook. Your investment in a prospect escalates sharply after he becomes an employee, so make your hiring decisions with great care.

Develop an interview form that covers the areas relevant to the jobs for the different types of employees you need to hire. Look beyond the resume to determine if the prospect really has what it takes to get the job done. Resumes can often look bigger than life, but the applicant can be ineffective when required to do independent, original thinking. Have the prospect describe in detail how he or she performed certain assignments in his previous job. This is particularly effective with engineering recruits. If they can't describe in detail how they solved a problem, chances are they either didn't really do the work or didn't understand the assignment. Be particularly careful when interviewing salespeople; they are trained in selling and may do a better job selling themselves than selling your products. One company president used to ask prospective salespeople to sell him a paper clip to see if the person was good at thinking on his or her feet and painting pictures for prospects. If you do not have good interview skills in your company, consider hiring a human resources consultant to help develop them and assist in the interviewing process.

The prospect's references should be carefully checked. One company almost hired a marketing administrator that claimed to have a business degree from the University of California at Berkeley. There was no record of attendance. Business references, other than employment verification, are difficult to check. Most companies or individuals will only provide a bare minimum of information because of the threat of lawsuits from disgruntled employees.

Personnel Motivation and Administration

If you recruit capable people, you must give them an opportunity to make a contribution. This means delegation of responsibility—one of the harder things for gifted entrepreneurs to do. Many potentially successful companies never get to the adolescent stage because the founder will not delegate authority and make the transition to a more professionally managed company. This transition can be facilitated by insisting that each manager take responsibility for his respective area even though it is possible for the CEO to micromanage the company during the early years.

Job descriptions can be used to motivate employees, although they are often considered to be large company tools for pigeon-holing employees. However, in order to hire a new employee, it's necessary to have a job opening, i.e., a job description. You should take the time to develop brief but complete job descriptions for all of your employees. An employee may have many different responsibilities—some of them unrelated—when the company is in its formative stages, but taking the time to document them can avoid many misunderstandings and foster a better sense of ownership. Job descriptions should be periodically revised to keep up with the company's growth.

Motivation is an important ingredient to maintaining high employee morale. Clearly established goals and responsibilities will empower your employees to do their job with pride. Listening to employees and acting on their suggestions is another powerful way to demonstrate your interest in their contribution to the company. Individual awards for exceptional performance should be considered as part of your motivational program. For example, the "Golden Banana Award" was established at TRW in its early years when an engineer came into the president's office with a new discovery and the president rewarded him with the only thing he could quickly lay his hands on—a banana. If you provide awards in the form of trips, consider including the winner's spouse or guest to encourage family harmony.

Employee evaluations are also an important part of management's responsibility, even for the young growing company. It is important to provide continuous feedback to employees on their performance, but a formal annual or semi-annual review is also necessary for the personnel file. Evaluations should not be a surprise for employees if the proper guidance is being provided on a regular basis. The formal evaluation provides a forum for the manager and the employee to establish expectations, areas for improvement, and to acknowledge contributions made in the past period. Reviews should be done on a consistent basis. This is an area that many managers avoid. (Renn Zaphiropolous, the founder of Versatec, a Silicon Valley manufacturer of high-speed plotters, solved the problem of untimely evaluations by tying a manager's salary increases to up-to-date employee evaluations!)

As careful as you may be in your hiring process, some employees will not work out and must be terminated. Use the probation period to carefully evaluate an employee's progress. If not satisfactory, terminate him or her before it expires. Termination of employees is a very difficult process and must be approached in a professional manner. Termination should not come as a surprise to an employee. You should inform employees as soon as possible if there are problems with their performance. Deficiencies should be clearly stated and goals for improvement clearly defined. Documentation of the situation is very important to be fair to the employee and to avoid litigation.

Once you have determined that an employee must be terminated, review the legal and emotional implications for your organization. If satisfied with your position, proceed quickly to complete the termination. Meet privately with the employee for about 15 to 20 minutes, explain the reason for the termination and let the employee speak to the issue. However, don't get into a debate over the firing. After the meeting, the personnel department (this may be you) should give the employee an exit interview and explain all termination benefits. A good checklist will help facilitate the exit interview and assure that nothing is overlooked. All company property should be collected and the employee should be supervised when he cleans out his desk. This should be done on the last day of employment. Care should be taken to protect valuable company property such as customer lists and manufacturing costs. For example, Handar terminated an international sales manager and did not take these precautions. The employee took every piece of information of strategic value he could lay his hands on and got employment with a competitor in the same position.

Fortunately, employee termination is not an issue that an entrepreneur faces every day. You can reduce your exposure to this unpleasant experience by focusing attention on recruiting good people and building employee morale. This is a tough job, and you need motivational tools to help you. Let's see how you can use the employee benefit package to build morale and commitment.

Employee Benefits Are Good for You, Too

How can you attract the best people to your company? Successful start-ups recognize that to recruit the top people in their field, the company not only has to offer an opportunity for contribution and advancement, they also have to treat their employees better than established companies. They can't offer much security (they've just opened their doors for business!) so opportunity, working conditions, business ownership, and benefits are the tools they have to work with.

These factors play different roles for the people you will be hiring. Most professional people change jobs to get a position that offers more opportunity for advancement and contribution—you don't want those that change just because of money. New college graduates, particularly technical types, are usually interested in the type of projects they will be working on and the opportunity to apply what they have learned. Direct labor employees are predominantly interested in wages, benefits, security, and working conditions. The compensation and benefit plans you develop for your company should appeal to the diverse interests.

Compensation is the stock options (particularly for start-up or high-growth companies), wages, and bonus package that a prospective employee receives to induce him or her to join your company. Insurance, vacation, and paid holidays are the benefits that are available to all employees. The founders or top management should set objectives for compensation and benefits for the company and develop a plan to implement the objectives. Industry surveys from the American Electronic Association (AEA) can help you compare your plan to industry practice. A plan is important to avoid major dislocations in your company's salary and benefit structure and to assure that you are competitive in the job market with other companies. Let's take a closer look at the benefits available to a start-up company and how they can be used.

Stock Options

Stock options are the most powerful tool you have to attract talented, highly motivated people to your company. It is the one compensation that a large, slower-growing company cannot provide to its employees. Where else can employees make ten times their annual wage in a lump sum distribution than in a stock holding of a successful start-up company? Many start-ups use this incentive as a strong bargaining chip to hire new professional employees—it's the one component of an offer that the current employer can't counter. Many start-ups try to hire at the same salary as the prospect is currently making and use the option as a deal sweetener to ferret out the ambitious prospects.

Financial Compensation

Salaries are the largest expense for most companies so a plan to manage them makes good economic sense. The plan should address all positions to assure that major dislocations don't occur between jobs in different departments that have the same value to the company. A job equivalent rating system should be developed so a production engineer can have the same pay scale as a development engineer, if the contribution to the company is deemed equivalent. The plan should be designed to pay for performance; the practice of automatically giving periodic pay raises, particularly for higher paid professionals, is becoming obsolete.

Compensation plans typically fall into two categories: one with significant incentives for higher paid professionals, and one that is directed toward the line employee. Both plans should provide incentives for performance. The plan for top professional employees should include a base salary, an incentive stock option, and an optional bonus plan. Bonus plans are often related to company growth and profitability. Trimble Navigation had bonus plans for top managers that did not kick in until after the company had exceeded 30% growth for the year. The president took responsibility for the company's profitability. Many companies construct salary plans for their higher-paid professionals with a fixed base, a component related to the company's performance, and a component related to achievement of negotiated goals.

The compensation plan for line employees should also contain an incentive for performance, particularly team performance. Cash profit sharing is successfully used by many companies. The company sets aside a fixed percentage of pretax profits in a profit-sharing pool. Twice a year employees are paid a bonus based on their wages as a percentage of all wages paid by the company. When the plan was first developed at Hewlett-Packard, profit sharing was paid weekly! Profit sharing rewards performance and provides an opportunity for employees to save money or make major purchases as a result of the company's success.

Compensation for sales employees should be aggressively set on the commission side. Sales employees are often the highest paid in the company; their commission should be 25% to 50% of total compensation. Commission structures should be progressive by rewarding higher performance at a higher rate. Many companies establish negative structures, which encourage sales people to go to the beach after they have reached quota.

Job offer letters should be carefully constructed; they are legal contracts that bind the company to financial commitments. The letter should clearly state the company's offer and financial commitments in case of termination. Promises for future advancement or pay increases should be avoided, because they can become legal commitments if the employee becomes dissatisfied or is terminated.

Basic Medical

Core medical coverage usually includes major medical insurance, disability insurance, and life insurance. Major medical costs emerged in the 1980s as one of business's big financial headaches as costs escalated at 15% to 20% per year. The problems for small business loom even larger in the 1990s since many insurance companies withdrew from the small company market in the 1980s. The effects of these changes can be mitigated by adopting a philosophy of cost containment, cost sharing, and insurance for catastrophic illness or injury. Affordable health risk will vary between groups of different economic standing; this should be considered when formulating the plan. But remember that benefits are easier to improve than to reduce. For example, many companies provide free insurance for their employees, but the employee pays for dependent coverage.

Major medical coverage can be provided in three ways: standard policy from an insurance company that allows the employee to visit the doctor and hospital of their choice, preferred provider organization (PPO), and health maintenance organizations (HMO). Standard policies cover major medical expenses with a deductible amount and co-insurance of 20% up to a ceiling of $5000 to $10,000. Cost can be contained by increasing the deductible from $100, which was common in the 1980s, to $250 or $500, which can save 10% to 15% of the premium cost. Deductibles were originally set at one week's wages when employee insurance was first widely adopted as a company benefit. The ceiling for the amount paid by the employees for a given incident can also be raised to reduce cost. A nonsmoking employer policy is another way to control costs, since some carriers provide nonsmoking plans.

PPOs are less expensive and provide the same basic coverage as standard policy plans, but the employee must usually use doctors and facilities that are members of the PPO. These doctors and facilities have agreed to provide services based on an established fee schedule. If the

PPO is a "swing plan," the employee has the option of using any doctor or hospital.

HMOs are another alternative for providing health care that became very popular in the 1980s; the largest is the Kaiser Permanente HMO. HMOs provide all services for a fixed fee, usually $5 to $10 per visit (even for a major operation) and are available to companies with more than five employees. They provide a form of institutionalized health care at competitive prices, but without the personalized service of the other two choices. HMOs are particularly attractive if the company has a lot of employees with small children.

Disability insurance is often overlooked by many smaller companies and employees of smaller companies, but the fact is that many more people become disabled and can't work than die. Disability insurance is relatively inexpensive compared to major medical, and offers additional security to your employees.

Life insurance is almost universally offered as part of company benefit plans and is quite inexpensive. The amount of insurance for an employee is usually related to their annual salary. Other medical benefits such as dental and orthodontic are not nearly so critical for maintaining the well being of your employees, and are usually found only at larger or more mature companies. These benefits are popular with employees, but the cost to provide them must be carefully weighed against the protection that is provided.

The medical benefit area is extremely complex and constantly changing. Be sure to work only with a qualified broker to set up your plan. Insist on references and check them out; there are a lot of incompetent people in the field. Once the plan is developed, you should select a viable insurance carrier that can meet your requirements at a fair cost. Make sure that the carrier is financially responsible and that the claims administration is done in an efficient and timely manner.

Retirement Plans

This is another area that is more commonly associated with larger companies and is seldom a high-priority item for the start-up company. However, after a company is established and profitable, some form of tax-sheltered employee savings plan should be developed, both out of concern for your employees and to minimize turnover. The impact on corporate cash is not as severe as it might first appear, because most plans are funded by pretax dollars. These plans often favor higher-paid employees and can be a good source of tax-sheltered savings for the owners of a closely held corporation.

Plans can take four different forms: a simplified employee pension plan (SEP), qualified profit sharing, 401K, and a defined benefit pension. All of these plans must be registered with the Internal Revenue Service and must conform to the guidelines outlined in the Employee Retirement Income Security Act (ERISA).

A SEP is a special IRA that can be established and funded by the company with pretax dollars. Up to 15% of employee compensation, to a maximum of $30,000, can be contributed for each employee. The reporting and disclosure requirements are simple, which keep down the cost of plan administration. Participants must be immediately 100% vested in the plan and are also eligible to contribute to their personal IRA. The employer contribution can be adjusted annually without any plan amendments.

The qualified profit sharing plan allows the company to contribute a percentage of pretax company profits to the plan. The plan can be established as either a defined or discretionary contribution. The discretionary contribution plan is the most popular because it provides more flexibility for the board of directors to fund the plan. The board can elect to make no contribution if the company has a loss or faces some difficult business conditions. The company's contribution is divided up between the employees according to salary levels. The qualified profit-sharing plan is popular among smaller companies because the administration is relatively easy.

The 401K plan is similar to the profit-sharing plan, but the company matches the employee's contribution. This plan encourages employees to save some of their earnings for retirement. There are participation tests that must be met so the plan provides less tax shelter opportunity for the higher paid employees compared to a qualified profit sharing plan.

The pension plan provides a defined benefit retirement income for the participants and is funded entirely by the company. This plan is the most difficult to administer because the investment portfolio must be maintained at a level to meet the plan's obligations to its participants. It is usually only found at well-established large companies and is not recommended for high-tech companies that compete in highly volatile markets.

The adoption of a retirement plan for your employees is a serious decision. The rules that govern the administration of these plans are defined under ERISA and are very stringent. The trustees of the plan, usually members of top management, can be punished by the Department of Labor if they do not meet their fiduciary responsibility in the plan administration. Consequently, an investment policy statement should be adopted that outlines your plan's investment goals and objectives. The policy statement should include the following:

▼ *Why the plan was established.* For example, it was created to help employees plan for their retirement and to hire and retain valued employees.

▼ *The investment objectives of the plan.* This should define the level of risk the plan is willing to assume in its investments and the rate of return it expects to achieve within the limits of risk.

▼ *Guidelines for the administration of the plan.* The guidelines should help the trustee delegate the administration of the plan and establish investment objectives and management rules.

▼ *Benchmarks for the performance of the plan.* The investment portfolio should be diversified and its performance tracked at least quarterly and compared to the performance of similar portfolios of indices such as the Standard & Poors 500 or the Shearson Lehman bond index.

▼ *Communication and reporting standards.* Procedures should be established for the trustees to communicate with the investment manager, the plan administrator, and the plan participants. All communications should be written.

The development of a form of retirement plan is complex, but it can become an important part of your benefit package as your company grows. The material presented in this section is only intended to familiarize you with some of the alternatives that are available to implement a plan. You should consult your attorney or a qualified financial planner with trusted references to help you develop a plan for your company.

Time Off

This is a significant benefit that is often taken for granted by employees. Most companies provide two weeks of vacation, ten holidays, and one week of sick leave which adds up to 10% of the work year. Your sick leave policy should be designed to encourage employees to bank sick leave for the time that it is needed, because non-exempt employees at most companies treat sick leave as vacation and use the time as it becomes available. Your policies should be constructed so that both sick leave and vacation time is earned as a percentage of the time worked. For example, the vacation accumulation rate for an annual total of two weeks could be 1/26th week per week worked.

Some companies set up a plan where vacation is accounted for on an earned and available basis. At the beginning of the year, the company credits the employee's account with two weeks of vacation. The employee is eligible to take the vacation before it is earned and pay it back during the year. Sick leave is usually only available on an earned basis.

Most companies also have a policy of awarding more vacation time for employees with longer service. For example, a schedule is established that increases the employee's benefit from two weeks at initial employment to six weeks after ten years of employment. This is another benefit that encourages valued employees to stay with the company; of course, it can also contribute to accumulation of deadwood.

Maternity leave, jury duty, military duty, funerals, and severance pay are other benefits that should be considered as you develop your plan. These situations will arise and are more easily dealt with if a policy has been previously established. By setting guidelines early in the company's life you will eliminate uncertainty when the issues surface and provide a basis for prospective employees to evaluate the company. Key prospects may reject your company if they perceive that you are disorganized or arbitrary in this area.

Other Perks

Companies have often provided other benefits to key employees such as luxury company cars, executive insurance, financial planning, and additional health benefits. Most of these benefits are now taxable under 1990 Internal Revenue Service rules and are not worth the trouble to administer. They also create a "ruling class" attitude within the company and can erect barriers between management and labor. We recommend that you avoid these in your company.

Besides the tangible benefits such as health care and retirement plans, there are many intangible benefits that can boost morale and improve working relationships. The Friday beer bust is legendary at many pioneering Silicon Valley companies such as Hewlett-Packard, its reputed founder. Today there are more legal difficulties with providing alcoholic beverages at company functions, but the idea can be embodied with nonalcoholic beverages. One of the most popular events in Silicon Valley is the Halloween party, when even the top executives come to work in costume to compete for the company's prize for "the best-dressed employee." It is important to include immediate families in these functions. High-growth high-tech companies require large time commitments from their employees; an understanding partner can mitigate much of the tension that could otherwise develop in personal relationships. Taking time to play together at regular functions and celebrating significant events, such as a large order, shipment, or significant product introduction, can provide a bonding between employees that will pay large dividends when times are tough and extra commitment is asked of your employees.

So how much do all these benefits cost? The cost can range from 30% to 50% of direct salary, depending on the benefits offered. The cost should definitely be included in your financial estimates for the business plan and in annual budgets. The objective of a benefit plan is to provide a level of security for your employees to encourage them to stay with the company and allow them to concentrate on their jobs. Benefit plans are a basic cost of doing business.

Benefit plans should be carefully communicated to your employees. Develop an employee handbook that explains the company's policies and benefits in detail and communicate the plan to prospective employees in a one-to-one meeting. Carefully explain new policies to your employees in communication meetings that are developed and attended by the plan administrator. A letter or personal address by the president should always accompany the issuance of a new benefits plan. Be sure to link the plan to the company's success in every possible way to provide maximum motivation to your employees.

In summary, the compensation and benefits area is extremely complex and a significant cost factor for most businesses. This chapter is only intended to provide limited background information to help you understand some of the issues. We recommend you seek expert assistance as you develop the plans for your company. There are consulting organizations and individuals that specialize in this area. They can help you develop compensation and benefit plans for your employees. When you receive advice from these experts, don't forget that the benefit package is your own. The form of this package is one of the key ingredients of the culture you will be creating in your company.

Corporate Culture Shouldn't Be an Accident

Every company has a culture all its own. You can sense it just walking in the door. Your company, too, will have a culture, whether the founders plan it or not. But you will be much better off if you plan your culture rather than just let it happen.

Company culture is a subtle thing. It is the set of values shared by the employees of the company, which arise from the philosophy of the company's managers. It is therefore imperative that you and your cofounders discuss values as you plan your venture. Just how deep is your customer orientation? What's your position on product quality? Prompt payment? The list goes on.

It is equally important that you define and communicate your set of values to your employees, no matter how many or how few you have. Why? So that all the employees will be moving in the same direction when problems arise or conditions change. This communication must be deliberate. You and your cofounders should be highly visible to the troops, revealing your values both by example and by more formal means like monthly meetings.

Among other talents, the CEO of a start-up company has to be a preacher, constantly articulating his or her values to the employees. This is a tough job, considering that the CEO also has to control a fast-moving organization and participate in hiring, firing, and other difficult personnel tasks. It's difficult to maintain a vision of your values when dealing with day-to-day business and personnel problems. As you step into this role, remember that actions define your values. Strive for consistency between your words and your deeds.

An excellent example of start-up culture was set by Tandem Computer. Founder Jim Treybig made a deliberate effort to form a culture based on honesty: He went out of his way to describe the long working hours to potential employees, and he quickly squashed gossip and office politics whenever they appeared in the organization. He saw to it that employees learned how the different parts of the organization were interdependent. And he established a two-day seminar on the company's culture, to be attended by every employee. Tandem also made a concerted effort to involve employees' families in the company. Spouses were included in the hiring process, invited to the company's Friday afternoon beer parties, and invited to attend the company's two-day culture seminar. It's hard to imagine how any company could do a better job of communicating values.

Unfortunately, there are examples of destructive cultures to report. One Silicon Valley company in financial difficulty was taken over by an investor. This individual, focusing solely on financial objectives, made it clear that the operative cultural statement was "No excuses—either achieve the results we have set or you're fired." He had invented a new style: management by fear (not by objectives, by communication, by walking around, or any of the other variations). Is it any wonder that the company lost its best people and went under? We could cite other examples as well, such as the start-up that overemphasized secrecy to the point that no one in the company knew what anyone else was doing.

Once your culture is defined, every employee must buy into it, or problems will arise. When people at all levels work to implement the same set of values, strong organizations emerge. But when one individual fights the culture, something has to give. In one start-up company the sales vice president continued to land unique contracts that were not in line with the president's more conservative philosophy, which drove the rest of the company. Eventually the situation came to a head and the sales vice president resigned. Afterward, the organization thrived.

Evidence of a company's culture is everywhere. Inside the company, the employee benefits package says a lot about philosophy. The form, content, and frequency of communication with employees is a mirror of the company's culture. Also, management practices like performance reviews are good indicators of culture. (Remember our example of Versatec, where supervisors were not eligible for pay raises if they were behind in their performance reviews? That's a cultural statement with real clout!)

From the outside, one can observe a company's involvement in community affairs or industry-wide activities. The company's promotional materials reveal much about culture. Perhaps the best indicator of all is a company's reaction to a customer inquiry or complaint, which tests the company's commitment to its customers.

Different approaches to these issues are neither good nor bad, just different. There is room for many philosophies in an industry. The important principles are that your culture must be positive, not destructive, and that it must be well communicated.

Your company will have a unique culture that flows from your business philosophy. Ask yourself, "What is my business philosophy? What values do I hold?" Once you understand your own set of values, you can then communicate them effectively to your employees. Your company will be much stronger for it.

CHAPTER 6

Marketing, Sales, and Support

The marketplace for high-tech hardware is a competitive battle ground. Rarely does a start-up company have the luxury of entering a field with no competition. Instead, you will encounter a chaotic situation, surrounded by older companies fighting to maintain their positions and grow. To survive in this warlike environment, you need a battle plan.

The front-line troops in your organization are your sales force. In order for these troops to be effective, they need a plan of attack. You have to give them direction on how to approach the marketplace. This is the job of your marketing organization.

There are two major elements of military planning: intelligence (about the terrain, the weather, and enemy troop disposition) and tactical plans (which units will move where to accomplish what missions). The same is true of marketing planning. You need good intelligence gathering, in the form of market research, competitive analysis, and trial marketing. And you need good tactical plans, such as new product introductions, sales forecasts, sales strategy, promotion, and the development of applications for your products. Your troops on the front line will have the greatest impact with these elements in place.

Your marketing people are your experts on the marketplace, continuously in touch with customers and gathering valuable information. They use this understanding of the marketplace to prepare the way for effective sales efforts and new product development.

A great example of a start-up company that stays in touch with its marketplace is Photon, Inc., in Silicon Valley. Photon is the manufacturer of the "BeamScan," an electro-optic instrument that measures the intensity profile of laser beams. Founder John Fleischer, although he did not have a marketing background, knew instinctively that staying in touch with his customers was the right thing to do. He made a point to travel around the country every so often, visiting customers whether they were planning to buy another instrument or not. As a result of this practice, Photon has an exceptionally loyal base of customers who have resisted inroads by would-be competitors.

Contrast Photon's experience with that of another small manufacturer of electronic test equipment. The product involved was a sophisticated computer-based system, which included proprietary software and hardware "bundled" to provide a unique solution to certain measurements. It was well accepted by users due to its high performance and reasonable cost. But by the early 1980s the PC revolution was changing the world of computer-based systems from bundled systems to open architectures with unbundled software. The founder of this company did not stay in touch with his customers during that period. If he had, they would have demanded the unbundling of his software and the use of a standard controller platform. As you might expect, orders for his otherwise attractive system dried up in favor of the more flexible, open alternatives offered by his competitors.

Since your marketing people become experts in both your products and their applications, you will be tempted to ask them to provide customer support: "Joe, can you take this call from a customer in Seattle? She has a tough question that I can't handle." Customer support is very important for any high-tech company. But it is equally important to insulate your marketing people from sales-related interruptions, and allow them to devote quality time to strategy, promotion, and other important tasks. If you don't do this, the strategic work just won't get done. Your authors can testify from their experience, in companies both large and small, that insulating a few key marketing people from day-to-day pressures is essential to long-term business success. Your customer support function, in contrast, should be fashioned to deal effectively with these very problems.

Getting the marketing job done well is critically important during your company's start-up phase. You need a strong, experienced marketing vice-president in the company at the outset. Sometimes this job can be filled temporarily by the president. But a full-time marketing person is required at least six months before the introduction of your first product. If the president wears the marketing hat, he or she must be able to perform all the marketing tasks with ease since he will have many other pressures. Let's examine just what these tasks are.

Market Research

We discussed in Chapter 2 the tremendous importance of verifying your product idea through basic market research. But don't forget that, after the planning for your first product is complete, market research has to continue. When the product begins to be used by customers, record their reactions and understand their experiences. You will soon be planning your second product, and you need all the market feedback you can get. Subscribe to the magazines your customers read. Get on the mailing lists of market research firms serving your industry. Collect all the tidbits of data on competitors or the market that come your way. Sales people, marketing people, and just about everybody else should constantly be on the lookout for useful information about the marketplace. For the high-tech company, staying in touch with the market is a way of life.

Competitive Analysis

No matter what your market niche, you're going to have competition. It may be direct competition from a similar product, indirect competition from other products, or merely alternative techniques available to the customer. In any case, it's important to understand the competition thoroughly before founding your enterprise.

When assessing a competitor, it is not enough to examine just the obvious areas of product performance, price, service, quality reputation, and the like. You must understand the capability of the competitor to introduce new products competitive with yours. You cannot anticipate what competitors will do. In fact, it's unwise to make predictions; they can get you into big trouble. But you can assess what they are capable of doing. A good model is that of Newport Corporation, a manufacturer of electro-optical products in Fountain Valley, California. Newport gained access to a technological development which would have allowed them to enter a large market. Before investing in the idea, Newport did some research to learn more about the new market. They learned from this research that the market was dominated by a single player, a major electronics company, and that Newport's competitive advantage would be fragile. Newport decided not to enter the market. They could not assess the competitor's intentions, but they knew that this company could do them grave damage if it so decided. Newport's decision was a wise one.

The story of Quanta-Ray, the laser manufacturer, was not so happy. In 1982, Quanta-Ray enjoyed the leading position in its market, having three competitors with lesser reputations. That year, the number three company, Quantel, introduced a new generation product that looked better than Quanta-Ray's on paper. The management team at Quanta-Ray discounted the threat due to Quantel's poor reputation. What they failed to recognize was Quantel's new and vastly improved product development capability. Quanta-Ray ignored the Quantel threat and concentrated on other things. A year later, they regretted their inaction as Quantel began to take serious market share from them. Now there's a lesson one marketing manager won't forget!

New Product Definition

It's hard to think of an activity in your start-up company more important than defining new products. New products are the lifeblood of any high-tech manufacturing company. Such a crucial activity deserves the utmost care from your organization, so take your time and do it well.

As we discussed earlier, market research aimed at defining your initial product is the first major activity of your venture. When you're talking with potential customers about your product idea, carry a list of its key features and get feedback on each of them. You'd be surprised at how easily a product definition falls out of such an effort. This principle holds true for indirect interviewing techniques like telephone or written questionnaire.

Defining new products is a team effort. The founders constitute the team for defining your first product. For subsequent products, there should be representatives of marketing, engineering, and manufacturing on the team.

The marketing team member brings to the table the results of his or her market research and competitive analysis. This person must be the expert on pricing conditions in the marketplace. Team members will look to him or her for input on distribution channels, since a product's design is often tailored to its intended channel. They will also ask this person for perspective on international issues, such as line voltage and frequency in foreign countries.

The engineering team member will be expected to know the state of the art in relevant technologies and be the team expert on estimating development cost. The manufacturing team member should know the company's planned manufacturing capabilities intimately. The team will look to him or her for estimates of manufacturing cost.

Together, the team members develop requirements for the new product—performance specifications, manufacturing cost, failure rate, etc. They also perform an analysis of the product's return on investment.

The result of the team's analysis and brainstorming will be a detailed definition of the new product and a vision of its journey through manufacturing and distribution to the end user. All of this should be documented in a product proposal. (See Appendix B for an outline of this document.)

The new product's definition must pass all of the following tests:

▼   Does this product fit within your mission statement? If not, it should be dropped. (Or it might be wise to review your mission statement.)

▼   Does the product exploit your company's competitive advantage? If not, it must be creating its own new advantage, or you shouldn't be considering it.

▼   Can you write a unique positioning statement for the product? If not, it should be redefined. There may be no demand for it.

▼   Is the product a close fit to market needs, and to timing considerations in the marketplace? If not, it should be redefined.

▼   Does the product fit your distribution channel? If not, is an effective channel available to you?

▼   Does the product fit within your present product line and into the product strategy outlined in your business plan? If not, you must re-examine your plan.

This last point on product strategy deserves some elaboration. You may be tempted to pursue a new product idea which is in conflict with your current offering or which violates the overall plan. Such product proliferation must be avoided. *Unless you are prepared to change strategy, you must stick with the plan that garners the maximum amount of business with the minimum number of products.* Your product development resources are too valuable to waste.

Likewise, you should watch out for products that will not contribute to the company's growth or that will significantly increase overhead expenses like selling or manufacturing more than profit. For a start-up company, every new product must be a winner!

Trial Marketing

It is often wise to try out your new product on a few customers before launching it for sale. If your product is aimed at a broad market, you might offer it to a limited geographical segment of the market to test its acceptance. If it is accepted in your test market, then you can proceed with the expense of full production, marketing introduction, and sales. It is far better to discover a problem with the product before you commit to these major expenses. For this exercise, the wisest approach is to hire outside experts to help you conduct the test.

If your product addresses industrial markets, you may want to install it at a few test sites to find any problems with the design, especially if it is complicated or contains much software. An *alpha* test site is an arrangement wherein you install a prototype of the product with a customer and he helps you to redefine it to fit his needs. A *beta* test site is used later in the design process; the customer is asked to work with the product and find any problems with it.

Customers generally enjoy the role of test site. They are flattered to have been chosen. They gain advantages over their competitors through earlier access to the latest innovation. And they can purchase the equipment at a lower price, since test site agreements usually include this provision.

Trial marketing is time consuming and takes a great deal of care to do well. It should be used when you have concerns about your product or your knowledge of the market. Trial marketing has a hidden advantage, too—the sites you choose for the trials will often become your best customers, references, and advocates.

New Product Introduction

As a new product nears first shipments, it's the responsibility of the marketing department to "introduce" it to the marketplace. This process is fairly involved. We'll devote an upcoming section of this chapter ("The Product Splash") and Appendix D to this subject.

Sales Forecasting

Here's a question that stumps many an entrepreneur: How can I predict how many of these things we will sell?

The first thing you have to realize is that sales forecasts are imperfect. Forecasting is an art in which you develop the best estimate you can with the data you have. Consequently, forecasts should be reviewed regularly and updated with recent data. But the forecasts will still be imperfect despite your best efforts.

The best indicator of future orders for a product is past orders for the same product. In your start-up situation, you don't have the luxury of historical sales data, so you must resort to modeling. Modeling is simply quantifying your view of the environment into which your product will be sold. You can choose one of four approaches to modeling your product's situation:

▼ Draw upon your own experience with similar or closely related products and discern the differences between them and this one.

▼ Scale the sales of your product to the known sales of a related product.

▼ Prepare a volume-price curve based on known sales of competing products. (See Appendix C.)

▼ Develop a unique model from primary data, such as the number of potential customers and their purchasing habits. This technique requires many assumptions and can easily lead to erroneous conclusions, so use it carefully.

A good example of modeling by a start-up is that of Ferretec, a Silicon Valley microwave filter manufacturer. Ferretec's proposed product was a filter used to purify the output signal from microwave sweep generators. Since market data was not available on such filters, Ferretec chose to estimate their sales by scaling them to the sales of the appropriate sweep generators. To do this accurately they had to know sales numbers for sweep generators and the scaling factor. Data on sales of sweep generators were available from the market research firm Prime Data. Ferretec determined the scaling factor by interviewing a number of customers and extracting a pattern from customers' answers to the scaling question. This technique worked quite well for Ferretec's new venture.

When a product's sales history has been established (say, six months after its introduction), you should review the forecast periodically as part of your production planning routine. Responsibility for this review should be transferred to your sales organization. The review period can range from quarterly to daily.

The periodic sales forecast is an exceedingly important element in the financial management of your company, since it drives the manufacturing schedule. If the forecast is too low, you will have to quote long delivery times to your customers and risk losing business. If the forecast is too high, you will purchase too much inventory—a waste of precious cash. Worse yet, a high forecast could entice you to set budgets too high in fixed expense areas like marketing, product development, administration, and manufacturing overhead. The lesson? Keep sales forecasts realistic—but conservative!

Promotion

Promotion is a marketing responsibility even though it is more tactical than strategic in nature. Promotion pursues a sequence of goals—creating awareness of the product, educating potential customers on the product's use, and creating a preference for the product in customers' minds. Sometimes you may even ask for the order in your promotional material, especially for less expensive items. There are many ways to achieve these objectives, but they are all guided by one principle: All of your promotional efforts should be derived from your positioning, and they must be absolutely consistent with it.

Here is a brief description of the major types of promotion:

▼ *Liaison With Opinion Makers*
The most effective promotion for any company is word-of-mouth testimony. This type of promotion is especially appropriate for a start-up company. You need to get to know key users, experts, and media personalities in your business area. These people can be influenced by demonstrations, loans of equipment, visits to your company, consulting contracts, or just conversations. Their testimony is the single best way to communicate your positioning to your potential customer base.

▼ *Publicity*
The next most effective promotion is coverage by magazine and newspaper editors. Statements by these sources are far more credible than statements made by your company; readers will correctly assume that you are biased but editors are not. Publicity has another great attribute—it's free! You should try to get all the publicity you can by educating editors, preparing press releases, and conducting press conferences. You can also write articles to be published in the trade media. These articles can be attributed either to the author or, changed slightly, to a willing editor of the magazine.

▼ *Advertising*
Paid advertising is a poor third choice to word of mouth or publicity, but it is often the best means of reaching a broad audi-

ence. Don't attempt to prepare your own ads; leave that to the professionals. You will get much better leverage for each dollar spent. Also, remember to plan your response to ads. Typical responses include mailing literature, calling the customer on the phone, or having your sales representative call.

▼ *Seminars*

Your message will have greater impact delivered directly to the customer. Direct contact is especially appropriate for products with significant technical content or difficult operating procedures. But it is also much more expensive per contact. One effective method is applications seminars. Seminars offer a controlled environment to demonstrate your product in contrast to risky demos by sales reps. They can be conducted at customer sites, at neutral sites like hotel conference rooms, or in conjunction with trade shows or technical conferences. A conference site can be quite efficient, since the attendees are pre-qualified and probably represent a large geographical area. You can charge a modest fee for attendance if the applications con-tent of the seminar is significant and your sales pitch is subtle. Otherwise, it should be free.

▼ *Direct Mail*

This is a relatively inexpensive way to deliver your message right to the customer's "in box." The technique is efficient if you have a fairly complete list of potential customers. With lower-cost products, you can ask for the order in the mailer. Read more on this subject in "Who's Gonna Sell this Stuff?" later in this chapter.

▼ *Trade Shows*

Exhibiting products at a trade show is an excellent promotional activity for most high-tech companies. Choose your shows carefully to concentrate your time and money on the best few. Your little company can look as big and impressive as its established competitors merely by allocating some real money to the preparation of your exhibit. Of course, the more you display, the more your competitors can learn about your products. But that's fair—you get to spy on them, too!

▼ *Newsletters*

A regular mailing to your customer base goes a long way toward holding their loyalty. Newsletters can also be aimed at potential customers, but their impact is less than a direct mail piece.

▼ *Users' Groups*

An interesting variation on the opinion maker and newsletter themes is to form a users' group, which would meet independently of your company to discuss applications and techniques. These groups are best suited to complex systems, where users have a need to share information with others. A users' group is a fine means of post-sales support for your customers. It is also an ideal opportunity for low-key promotion.

▼ *Literature*

Good technical data and supportive literature are musts in high-tech industry. The more complex your product, the greater the need for written material to explain the product to the potential buyer. Each product should have a flagship item, typically a brochure or technical data sheet, to discuss its performance and applications. Common literature used in the industry includes company brochures (describing the company), product line brochures (describing several related products), technical data sheets (describing one product in detail), configuration or selection guides, price lists, and applications brochures. All of these items should reflect the positioning of your product, company, or both. Even your operating and service manuals can include low-key promotional messages.

Start-up companies should always prepare a simple company brochure to deal with the inevitable question, "Who are you?" If you're selling expensive capital equipment or OEM-type products, the company brochure is important for establishing credibility.

▼ *Videotape*

High-tech companies selectively use videotapes in lieu of literature for both consumer and industrial products. Videotapes are good visual presentations of complex applications, and they can spice up most sales presentations. At just a few dollars per copy, they cost no more than full-color brochures.

When it comes to promotion, a start-up company can't do everything. There simply isn't the time or money. So you have to be selective. Some product promotion match-ups seem natural; for instance, trade shows are an excellent means of promoting capital equipment that doesn't lend itself to written descriptions. You should choose your mix of promotional vehicles to include those most likely to appeal to customers. (This is another reason why you need an intimate understanding of your customer!)

Many of the tasks involved in promoting your company's products can be handled by your employees. But preparing an array of visual materials cannot. Your company should retain a marketing communications agency to prepare your literature, advertisements, direct mail pieces, and other promotional material. The professionals in these agencies will tell you how to get the most impact from your promotional dollar. Hire an agency six months before your first need for publication, and let them help you plan your campaigns in coordinated fashion. You may also find a public relations consultant useful in planning and implementing your publicity.

Promotion involves more than just contacting an agency. Your marketing people must work closely with your agency, providing guidelines and specific objectives for every campaign.

All the key players in your company should seek opportunities to get your message out into the world. Promotion may be a marketing function, but it's everybody's job.

Applications Development

High-tech hardware frequently has uses beyond what the inventor initially envisions. Your marketing organization should keep a collective eye open for such "incremental" applications. Once an attractive new application has been identified, the marketing department can use its arsenal of promotional devices, special modifications, or recommended product enhancements to capture the new niche.

Opportunities to develop new applications are of at least four types:

▼ *Expansion of a market niche that the product has "found" of its own inertia.* Appropriate marketing tools to use are focused promotion and focused sales efforts.

▼ *Filling a new market niche that is satisfied by your existing product but which you have not developed.* This situation calls for the type of promotion and sales efforts used for new product introductions described later in this chapter.

▼ *Correcting deficiencies in the product which are preventing sales into targeted niches.* This form of applications development is best accomplished by special modifications or product enhancements.

▼ *Filling a gap in your product line with a new product.* This situation can be addressed by an enhancement or a new product definition.

Sales Training

Since your marketing people know a great deal about both your new products and their markets, it's logical that they should train the sales organization on new products. The training should stress the intended applications as well as the operation of the product. Sales people also need to know the identity of the targeted customer and what benefits the customer might perceive in the product. A good sales training package will also include comparisons with competing products and an overall strategy for approaching customers.

It is best to record all of this advice in written form so your sales people can refer to it after the training session. Above all, be brutally honest in this material. It does a salesperson a disservice to allow him or her to walk into a competitive situation with an overly optimistic view of your product's advantages. An outline of a sales training seminar is contained in Appendix E to help you tailor your training manual.

Setting a Price is Like Walking a Tightrope

Price is always an issue when selling a product, regardless of how well trained your sales force is. The traditional pricing approach for innovative high-tech products is "Whatever the market will bear." But that's not quite true.

Pricing is really a balance of opposing forces. Customers, of course, want a low price. Competitors will push your prices down as you fight for market share. Yet all of your internal forces press for the highest price to maximize profit. These internal forces are just as valid as the external ones. Profitability is the key to your company's long-term survival.

So how do you determine the optimum price, which balances profit with market share? The key issue is *value*, the ratio of what the customer receives to what is paid for it. What your customer receives is characterized not only by technical performance but also by "softer" attributes such as user friendliness, warranty policy, service, cost of ownership, and post-sales support. All of these attributes need to be weighed and compared to the offerings of your competitors before you can set your price. In this weighing, don't forget to estimate any improvement your competitors are capable of achieving over the time frame it will take to bring your product to market.

Your customer may have access to alternative techniques to achieve the same end that your product addresses. You need to understand these competitive techniques, even if they don't involve hardware like yours. Another alternative for the customer is to "do nothing," that is, not buy any solution to the problem at hand. Just as with competing products, you must understand the price the customer would pay to acquire each of these alternatives.

You also need to look within your company to quantify the ramifications of a pricing decision. Estimate the manufacturing cost and the support costs of each of your products. Only then can you understand the sources of your company's profits. You'll find that this internal analysis of pricing decisions is relatively easy compared to external analysis.

Having gathered all these data, you are ready to set a price that will result in a predictable sales volume with a known profit margin. You can do this by estimating the trade-offs that will occur within customers' minds, as your product is offered at different prices. Be sure to analyze these volume-vs-price trade-offs from the viewpoint of the customer, not the viewpoint of your company. It's your customers who will make the purchase decision based on *their* perception of value.

How should you go about analyzing all these tradeoffs? We have found that the most convenient tool for most entrepreneurs is a volume-price graph as described in Appendix C. The resulting curve represents your forecast of sales as a function of price. Granted, this result is quite subjective. But it is an excellent vehicle for you to address all the issues surrounding pricing and the relationships between them.

With your volume-price graph in hand, you can make the pricing decisions necessary to achieve your strategies. Here are some examples of price-related strategic decisions:

▼ Price for a given market share

▼ Price for maximum profit dollars

▼ Price for the best trade-off between profitability and competitive posture with a specific competitor

Pricing is a big element of your positioning strategy. You can't pursue a low-cost commodity strategy if your price is higher than those of major competitors. Conversely, it's hard to convince customers that your product is the highest performance or the finest quality if it has the lowest price. The personal computer market that evolved at the end of the 1980s provides a good example of these principles. At the low end, Dell Computer achieved good market penetration as an IBM "clone" manufacturer with a positioning of lowest price with good quality. Their promotion focused almost entirely on price, which was correct for Dell. At the other end of the spectrum, Apple Computer carved out the high-performance/high-price niche. Their promotion focused on performance and usability, especially for desktop publishing (Apple's biggest market segment). Apple's personal computer products were consistently priced at the high end of the scale. The consistency of price and positioning was an important element in Apple's success.

Volume-price analysis assumes the existence of competing products with characteristics that are known to customers. But what if your product idea is revolutionary? It might have a much lower manufacturing cost than anything of its type, or it might represent a new approach with significant value for the customer. In the case of breakthroughs like this, you must study customer habits to understand the value they place on your product idea. Then you can price the product according to that value perception.

A common technique for estimating value is calculating the payback period for savings which industrial customers will experience as a result of owning your product. Hard economic benefits such as savings in labor, maintenance expenses, and consumables are very meaningful. And they're usually not difficult to quantify. If your customers believe they would save $10,000 per month by using your product, then they might be willing to pay $360,000 for it—a payback period of only three years.

The risk in breakthrough pricing is underestimating the competition. If your product can be copied or duplicated in some way, then you should assume that it will. A classic example of this problem is that of Hewlett-Packard's introduction of the hand-held scientific calculator in the early 1970s. Believing that the market was small, H-P set a high price, $395, for this unique device. But it turned out that the market was huge. Soon a number of competitors entered the market and took advantage of H-P's price "umbrella." The price of equivalent calculators soon fell to near $100, and H-P's market share eroded dramatically. This shows it is possible to price an innovative product too high.

Is it possible to price your new product too low? Absolutely! This mistake is made more often than you might think. Common causes are failing to fully understand the value the customer perceives in your product (another argument in favor of market research early in the game) and an incorrect estimate of manufacturing cost. One manufacturer of expensive capital equipment once landed a distribution contract which guaranteed a significant worldwide market share. Since the contract's volumes were larger than the company had experienced to date, they assumed that manufacturing costs would come down with volume and agreed to a low price. But the cost reductions didn't happen. Contractually obligated to deliver units at the low price, the company lost money for some time. They were eventually acquired by a competitor (presumably at a low price!) because their technology was excellent. This calamity could have been avoided if they had examined their cost assumptions with skepticism before making the pricing decision.

Much high-tech hardware is sold in conjunction with other hardware or software, from simple instrument clusters to complicated computer systems. Managers of these systems businesses face the haunting question— should we "bundle" a number of items together into a single package and single price? The temptation to use bundled prices is great. You can get greater dollars per sale by causing some customers to buy items they don't really need. Also, you can avoid the problem of determining whose fault it is when a failure occurs in the system. If your company has supplied all the components, then you implicitly take responsibility for the performance of the whole system.

Unfortunately, customers generally prefer unbundled systems. With an unbundled structure, they can choose whether they want each item in your list—for example, whether they need the display in a data acquisition system. And they can mix products from different vendors to optimize system performance for their applications, as in choosing the disk drive for a PC system. The typical customer does want you to take responsibility for system performance, however. In the unbundled case, you can offer it as an additional item.

Our advice concerning bundling: Ask yourself if the additional items are essential to the operation of the system. If the answer is "no," then don't bundle items together. If the answer is "yes," then it's okay to bundle the system. But be sure that you're not creating a negative incentive by denying customers options that they might want. This advice concerning bundling applies not only to hardware and software products, but also to the service and training that accompany your product.

The Product Splash

Pricing is certainly one of the most important decisions you will make when you introduce your product to your customers, but how you promote it will significantly influence the price you can charge. Every product needs a careful introduction to its marketplace. The greater the impact of the introduction, the faster the product's sales will rise to a mature level, and the more market share it will achieve. These principles apply to all high-tech products and to all companies, large and small. Your start-up company needs to execute new product introductions just as well as larger, established companies do—maybe even more so!

New product introductions are the best example of how a promotional campaign springs from a positioning statement. Your introduction will be much stronger when you develop the right positioning statement for your product and follow it consistently. (Remember the example of Compaq Computer in Chapter 2?) The key to successful positioning is the accuracy of your message *from the customer's point of view.*

Introducing a new product is a lot like a military operation. There is a great deal of work to be done, and it must be carefully coordinated. There must be one person in charge to direct the traffic. This person will normally be the product manager in the marketing department. He or she should be the head of a team of people from the functional groups within your company—product development, manufacturing, marketing, finance, and quality assurance. This team should meet regularly to iron out all the problems that inevitably arise. The biggest headaches in new product introduction will occur inside your organization, not out in the marketplace. The slippage of an engineering schedule or failure of a tooling design can have large repercussions on an introduction plan.

Yes, you need a new product introduction plan. Even though you may not enjoy writing, your plan must be committed to paper. First, it's a way to insure that everyone in your organization goes along with the plan. Secondly, it notifies everyone of their obligations surrounding the introduction. Finally, it forces the product manager to think of everything at the outset of the project, thus avoiding omissions or mistakes. The plan does not have to be long and involved. For simple products, it might be only one page long. See Appendix D for a plan outline.

You have the opportunity to exercise some creativity in preparing your introduction plan. One objective you seek is to gain the attention of prospective customers, and there are unlimited numbers of ways to do that. When NeXT, the Silicon Valley workstation manufacturer, introduced its first product, they did it with a spectacular press conference. They had a CEO, Steve Jobs, with considerable fame, and they took advantage of it. Everyone wondered what his next venture would be after leaving Apple Computer, and NeXT got a lot of attention as a result. When Hypres introduced its first instrument, an ultra-high-speed sampling oscilloscope using superconducting circuitry, they did it at a trade show. Hypres arranged to have a modest booth against a wall. They hung a huge sign on the wall that said SUPERCONDUCTING OSCILLOSCOPE. Their booth was jammed!

Remember that there are other ways of promoting your product besides the usual magazines, trade shows, and seminars. Other opinion-makers in your market might be valuable. How about promin researchers or influential customers getting a "sneak preview"? Such unique approaches to promotion often turn out to be the most powerful.

It is handy to summarize the events of a new product introduction in a timeline chart. It gives you an easy-to-understand presentation of how events relate to each other. Let's look at the sample chart in Figure 6-1. Your chart might be more or less complicated, depending on the product. Prominent items included in the chart are:

▼ *Availability*
An introduction should be timed according to when the product will be available to customers. There are many reasons why companies are tempted to pre-introduce a product—competitive pressure, an upcoming trade show—and every one of them is wrong. When a product is introduced, the customer should be offered three things: acceptable specifications, an acceptable price, and an acceptable delivery. Customers are very unforgiving when you fail to deliver all three. So begin your planning with the question "When will we be prepared to deliver?" It will save a lot of headaches later on.

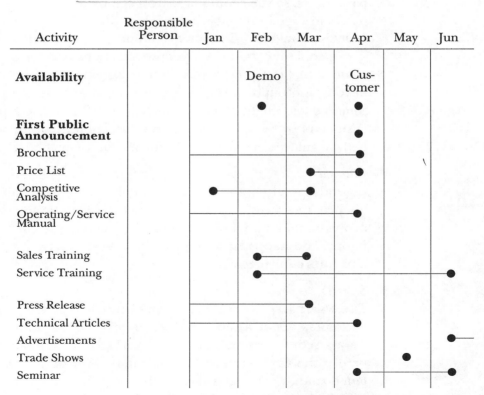

Figure 6-1: There are numerous promotion and marketing tasks required to be done within a relatively short period of time when a new product is introduced.

▼ *First Public Announcement*

Your plan should state a specific date when the public will hear about the new product. Sometimes there are different dates for different types of customers, or, more commonly, for geographic market segments.

▼ *Sales Literature*

Publications such as brochures and price lists must be ready to hand out on the day of the first public announcement. Customers will be impressed when their inquiries are answered with a professional-looking document. If in-depth technical information is necessary to fully explain the product's performance, it should also be available at the same time. Other literature, such as applications notes, also have an impact if they're readily available. But often you don't have the resources to do everything at once, so these items get spread out.

▼ *Competitive Analysis*

Customers demand to know how your product compares to competitive offerings, sometimes in a number of performance parameters, so you should prepare a complete analysis and be as objective as you can. Objectivity is key—don't fool yourself! The analysis should be distributed to your internal marketing and sales people for use in competitive situations and for planning promotional campaigns. You should also give it to professionals in your distribution channel. You may need to be selective here; some people might allow the analysis to fall into the hands of customers or competitors, which is bad news. Note that competitive analyses are also very useful in strategic planning.

▼ *Manuals*

A good operating and service manual is extremely important for just about any high-tech product. Delivering a complete, professional manual to customers is a sure way to make them happy, and happy customers are your best source of referrals and future business.

▼ *Sales Training*

The worst surprise you can spring on a salesperson is to tell customers about a new product before you tell him or her. Always schedule some form of training for your sales force before the first public announcement. Even if you plan detailed training at a later date, don't neglect to give your salespeople the tools and information they need to answer the first questions that customers will throw at them.

▼ *Service Training*

Training of your service people or outside service organizations should occur just before or immediately after the first public announcement. Don't schedule it too soon or they might forget most of it by the time they perform their first repairs.

▼ *Promotion*

The first rule of new product promotion is get your free publicity before you start to place paid advertising. As we mentioned earlier, editorial coverage has good impact because of its credibility. To get to the editors, you can hold a press conference if your new product is a significant breakthrough in a large market. Otherwise, your best bet is to prepare interesting materials and visit all the editors in your industry or arrange for them to visit your company, which is advantageous. A poor third choice is to send them press releases. Remember that editors are looking for news for their readers. If your product is newsworthy enough, maybe you can get the front cover of a trade magazine! Be sure to start your publicity effort early. Trade magazines and other sources of publicity often have long lead times— up to six months for a major story.

It is important to establish momentum in a new product introduction. Be sure to schedule other forms of promotion like customer seminars, visits to important customers, invitation-only demonstrations, and direct mailings soon after the first public announcement.

Note that the timeline in Figure 6-1 assumes the fortunate situation that the new product will be ready just before a trade show. It's great to announce a new product just before or perhaps during a major trade show, but beware: The attraction of a trade show can be a fatal one if it causes you to introduce your product before its time. Missed delivery promises and incorrect specifications or prices cause irreparable harm to young companies. Don't rush your product to meet artificial introduction deadlines. There will always be other trade shows.

Not every product deserves the full introduction treatment described here. The appropriate level of effort is determined by the complexity of the product, its importance to your company, and the characteristics of the marketplace. You should review all major aspects of the introduction process for each new product and prepare some kind of a plan, even if only a one-page memo. By following these guidelines, you can maximize the impact of a new product and get a good start on the return on your investment.

Who's Gonna Sell This Stuff?

After you have introduced your product to the marketplace, you need to get some sales. After all, you are in business to make a profit, and profits come only after customers have paid you for products shipped. And you can't ship unless you have an order. It all comes down to orders. The sales function is perhaps the most critical function of a high-tech company, but it often doesn't get the attention it deserves.

Creating a good sales organization is a big problem for many companies, especially high-tech start-ups. They often underestimate the difficulty of creating a good channel of distribution. One start-up we observed had a significant opportunity in instrumentation but a limited time window in which to exploit it. The founders were too busy with product development, finance, and other diversions to spend the time to set up the needed direct sales force. Although they had achieved good momentum with test sites and trade show exhibits, they lost it all by delaying the sales organization. They eventually built the organization, but not before the competition had nearly caught up with them. The time window closed, and the company never realized its potential.

It is possible for your sales organization to be a source of competitive advantage. Here are some examples of ways to do it:

▼ If there is a large company operating in the same market with non-competing products, you might convince them to recommend your product and work cooperatively with your sales people. You can then ride on the larger company's coat tails in many sales situations, and your competition will be shut out of a possible reference.

▼ If you can sign up the best sales representative firm in the business, your competitor will have to settle for less.

▼ If your product requires on-site installation and service, one option is to train the service representatives to be salespeople as well. In this way, every installation or service call is also a sales call! This technique was used by Quanta-Ray, the start-up laser manufacturer, and it worked exceptionally well. Quanta-Ray grew to the number one spot in its market niche without having to add a costly sales organization on top of the field service team.

So how do you decide what form of sales organization to use? You must first understand your customer, your product, and how they interact with each other.

Your customer can be characterized by geographical location, industry, department, and job function. Sometimes it is not obvious who makes the ultimate buying decision, so you might need to research this question. You should understand how the buying decision is made; for instance, some industries have elaborate approval procedures for the purchase of capital items. The length of the purchasing cycle, from on-the-spot decisions on PC peripherals to tortuous two-year procedures for big ticket items, can be a major factor in determining your choice of sales channel. You also need to know whether the customer requires a demonstration before buying.

The price and complexity of your product are important factors in determining the best distribution channel. Price (actually the size of the average sale) is a fundamental parameter which will shut out many potential avenues. To understand the complexity of your product consider questions like:

▼ Is it hard to learn to operate? Or trivial to learn?

▼ Does it have many components? Or just one?

▼   Is there software which must be understood and learned?

▼ Is the product customized for most orders?

Finally, your product's relation to other products on the market, especially those of your company, is a factor. Is this product an accessory to another? Is it a mainframe which will be surrounded by other accessory products? Are there other products which are closely related to yours in the marketplace?

Attributes to Look for When Choosing a Distribution Channel

Now that you have analyzed your product and market, you can work to identify the best type of distribution channel for your company. If you're lucky, the choice of channels may be easy—for instance, a dealer network for accessories which are purchased along with PCs. Then again, you may have tough decisions ahead.

Compare distribution channels with the following questions:

▼ *Does this channel reach my customer?*
You should insure that the channel has the required geographical coverage and is in touch with the right types of companies or individuals.

▼ *How effective is the customer interface?*
People working in your channel must have some knowledge of your product's application area. Also, their degree of cus-

tomer contact must be sufficient to satisfy customers' need for information.

▼ *Does the channel have the necessary level of technical support?*
You can provide the support from your own company, but you need to plan how this will be accomplished for each distribution channel.

▼ *What will be the cost of using this distribution channel?*
There are two sides to cost—setup costs arising from hiring, training, and preparing materials, and the cost of managing and supporting the channel. The distribution between fixed and variable cost components can be important, especially for a start-up hard pressed for cash and unable to support the fixed expense of a direct sales organization.

Some other questions are less critical but well worth looking into:

▼ How much control will your company have over the sales organization? You need control over every critically important element of your business.

▼ Does the channel have a high degree of personnel turnover, or is it relatively stable? The cost of retraining can be considerable.

▼ When using this channel, how close will your company be to the customer? Good customer contact is essential for quality feedback and new product ideas.

There are eight types of domestic U.S. distribution channels for high-tech hardware:

1. Direct sales force

2. Telemarketing

3. Sales representatives

4. Stocking distributors

5. Retail outlets

6. Direct mail

7. Catalogs

8. OEMs and VARs

Direct Sales Force
For some products, a personal interaction between your company and the customer is required before the customer will buy. This is

especially true when the product is expensive, complex, or customized. Many industrial products fall into this category.

One way to achieve direct contact is to have your own sales organization calling on the customer. With such a close customer relationship, quality feedback and market research information are much easier to come by. You also will have greater control over the following factors by using your own organization:

▼ What they say about your company and its products

▼ Which products they emphasize

▼ How well they adhere to policies on discounting, expediting, product returns, etc.

▼ What is said to them in performance reviews

▼ The selection of salespeople

The cost of a direct sales force ranges from 6% to 20% of sales. For most items, 10% is a good planning number, although complex items like computer systems average more like 15%. These numbers should be increased by a point or two for the required sales support organization in the factory. This relatively low cost is attractive. You must understand, however, that this cost is fixed and the percentage of sales it represents depends on whether the sales force is producing the needed volume. What's more, you will need to spend money up front to hire and train people.

For expensive industrial products, the direct sales force is usually your best bet. Even if you use another route in the early stage of the company, you should strive toward the eventual establishment of your own sales force.

[handwritten margin note: Not done on MM, with disastrous results]

Telemarketing

A telephone-based sales channel provides direct contact with the customer but lacks the impact of face-to-face discussion. You should consider this form of distribution channel when your average sale is below $5000, where direct sales forces are not cost-effective. Telemarketing is appropriate only when your product is fairly simple, so the customer may be willing to buy sight unseen after a brief discussion about the product. Examples of such lower-cost items are cables and and other accessories, simple test instruments, or PC peripherals.

The key to effective telemarketing is having a good list of prospects to call. For industrial products, you should have your own list. However, this can take years to create. As a start, you can purchase lists from professional associations, magazines, or trade organizations. You can begin to supplement these lists immediately by collecting inquiries from advertising, direct mail, and trade shows.

As with direct selling, you should have your own telemarketing group within the company, primarily for reasons of control. Competent telemarketing organizations are available, however, to satisfy your needs until you can afford your own. Their cost will be somewhat greater than an in-house group.

The cost of a telemarketing organization ranges from 5% to 10% of sales, of which one-half is the telephone bill. Clearly, this is an attractive alternative for those products which lend themselves to such a sale. But it is rare that telemarketing is the sole sales effort used for high-tech hardware. Most often, it is employed in conjunction with heavy advertising or direct mail.

Sales Representatives

This is a third form of direct selling. You might call it "rent a direct sales force." In this channel, you contract with a professional sales organization to sell your product for a commission. It can be quite attractive for product lines with average transactions over $5,000.

Sales representative organizations exist to serve many industries—computers, industrial controls, instruments, manufacturing equipment, and components, to name a few. They tend to be quite professional and experienced, and they never carry competitive product lines as some distribution channels do.

The "sales rep" channel will cost more than a direct sales force since they are profit-making organizations and they tend to pay their people well. For capital equipment, commissions run in the range 12%–15%. For components, commissions vary from 5% to 15%, depending on your volume. The attractive aspect to this cost is that it is completely variable—that is, you don't pay a commission until after the order is entered.

Since the cost of sales representatives is tied to orders received, it is possible to "rent" a lot of bodies to achieve wide geographical coverage. There is a limit to the number you should sign up, however, and that is the cost to train and manage all the people. This is a key point. Distribution partners, such as rep organizations, require considerable supervision on your part. The required internal sales support organization can easily cost 5% of revenue. As a rule, it is best to organize for complete geographical coverage of your markets using the least number of rep firms.

Stocking Distributors

Distributors are an excellent means of achieving wide geographical coverage for lower-priced industrial products, such as ICs and low-cost computer peripherals. They buy quantities of your product, put them in stock, and provide fast, local delivery to customers. Since they carry a variety of related items, they offer the customer the convenience of "one-stop shopping."

Distributors are far more than just warehouses. They have salespeople who call on major accounts in their local area. In some cases, the distinction between a distributor and a rep may be only that the distributor takes physical possession of the product. But the distributors' sales forces are different. They emphasize the business relationship between their company and the customer and often don't push individual products. They may also carry competing lines, so take care to choose your distributor carefully.

Distributors will require a discount from you, usually ranging from 15% to 30%. For planning purposes, 20% to 25% is a good range, although the discount is heavily dependent on volume. This discount may seem high compared to other distribution channels, but remember that distributors provide a large service to you—carrying the inventory required for stock delivery.

Retail Outlets

Retail outlets also provide a degree of personal customer interaction. If your market is consumers, small businesses, students, or other mass populations, retail outlets might be your best distribution channel. But be careful: The technical capabilities of retail salespeople vary widely, so you must feel confident that your product can be sold with relatively simple question-and-answer interaction.

Since the customer has to come to the store to buy, it is important to supplement this channel with heavy advertising, direct mail, or other promotion. Distribution contracts with retail outlets will often include joint promotional efforts.

Retail outlets will expect a discount from you ranging from 25% to 50%; for most high-tech products, plan on 30% to 35%. The initial cost of training and the ongoing cost of promotion will also be significant. In some industries, you may have to deal with the outlets through the reselling distributor, an intermediary who will typically require a 40% discount.

Retail distribution is expensive. Be sure that you can absorb these costs in your pricing.

Direct Mail

Direct mail is an alternative for those products which can be sold sight unseen. This sales channel can be quite powerful. You are delivering your message directly into the hands of the prospective customer, and you are in complete control of the message with no intermediate entity to filter it. It is appropriate for products of all types priced under $1000.

As with telemarketing, the key to successful direct mail marketing is the mailing list. You can build your own or hire a direct mail organization to do the mailings for you.

For simple products, you can ask for the order in the mailing, perhaps by including an order blank. This familiar technique is used in the merchandise offers that credit cardholders receive. (By the way, consider accepting credit card payment for lower-cost products offered to consumers and small businesses.)

If your product requires technical support during the sale, you need a means to provide it. One approach is by toll-free telephone.

Catalogs

Catalogs are a variation on the direct mail approach, wherein you hire a professional sales organization to sell your product for you through the mail. These catalogs carry combinations of items targeted at the same customers, and the mailing lists are carefully groomed to penetrate only one market. The catalogs are updated and mailed to the list periodically. A good example is "Personal Computing Tools," aimed at engineers using PCs. This catalog contains industrial PCs, PC-based instruments, computer-aided design packages, software, and various accessory products useful to the targeted audience.

Be careful if your product requires technical sales support. Most technical catalog companies do offer technical support of the items in their catalogs, but the quality of this support can be spotty. An alternative is to offer a toll-free number in the catalog whereby the customer can call your company directly for the answers he needs.

Most catalog companies work like distributors, buying the product at a discount and offering stock delivery. Brand names are often removed. Catalogs are expensive compared to other channels due to the large effort they make to sell the products. Discounts range from 25% to 50%, with 35% to 40% being average.

A catalog can be an excellent distribution channel for a relatively low-cost (up to a few thousand dollars) product with a well-defined market. Of course, a competent catalog has to exist in your market for you to be able to use this channel.

OEMs and VARs

Selling to original equipment manufacturers (OEMs) is different from other types of business transactions. This form of business can be profitable, but it requires a fundamental commitment to the peculiar requirements of the business as described in Chapter 3. The same is true for selling to value-added resellers (VARs), which are companies that buy high-tech products (typically a computer system), add value to them (typically software), and sell the resulting system to end-users.

A component manufacturer can use distributors or other channels to sell to OEMs, but an equipment manufacturer must sell direct. The organization to do this can be quite modest, and it shouldn't cost over 5% of sales.

Your relationship with an OEM or VAR should always be spelled out in a written contract. The contract should contain these major elements:

▼ Your customers stipulate that they are an OEM or VAR, not just a reseller of your product.

▼ You grant them a substantial discount from list price based on volume. This is to protect them from other channels carrying your product and to acknowledge their value to you in saving selling and promotion costs.

▼ The customers stipulate the number of units they plans to purchase and when. Their delivery requirements are described.

The discounts granted to OEMs and VARs depend greatly on volume, and range from 10% to 40%. In this case, however, the single unit price may be artificially high since all sales are done in quantity.

Should I Use a Direct Sales Force or Sales Representatives?

Since much high-tech hardware is relatively expensive capital equipment, this question arises frequently. The answer to the question must be uniquely yours. It depends on your company, your products, and your customers.

A direct sales force has several advantages due to the intimate contact between your company and the customer. These include improved quality feedback, more effective market research, and better brand identification by the customer. Fully utilized, it will be less expensive than reps. The people in the organization will be your choices, not someone else's, so you can count on having your desired image presented to the customer.

Another advantage of a direct sales force is your ability to control the organization. Company policies like commission splits and prospecting requirements are readily enforced. You can expect better forecasts from your own organization. And coordination of customer support—for example, solving incorrect shipments or handling cockpit error problems—is easier too.

A direct sales force has two disadvantages, both relating to money. First, it is expensive to hire, move, and train a group of professionals, and it takes time to do it. Second, the expenses associated with maintaining the organization are fixed, and only slightly related to sales results. So the decision to create a direct sales force carries financial risk with it. You will have to spend a lot of money first, hoping that your sales forecast is not overstated.

There are certain attributes of your product and your market which indicate the need for a direct sales force:

▼ If your average order size is quite large (for example, over $50,000), then a direct salesperson will be cost-effective.

▼ If your product's sales cycle is as long as one year, then external sales channels will be difficult to motivate. The most effective organization in this case is a direct sales force.

▼ If your product has relatively few customers, then a small direct force targeted at them can be cost-effective.

▼ If your product is complex, it may be impossible to train external channels sufficiently. In this case, you will wind up closing each sale yourself, so you might as well go direct.

▼ If your product is customized, you will be involved directly with each sale also.

▼ If pricing is volatile in your market, then you may need to participate in each sales situation. This is common in markets where each sale is preceded by an "invitation to bid," as in some government-oriented businesses.

Sales representatives offer the start-up company a significant advantage in that they are in touch with the market and ready to go at any time, so you can expect a faster start-up of your sales. They have less personnel turnover than direct sales forces, and they often have excellent mailing lists to use in promotional campaigns.

The greatest advantage of representatives is their variable cost. To the start-up company, this means less cash spent at the front end and less risk later if business doesn't grow as fast as forecast. It also implies that you can afford to hire a large number of representatives and thereby achieve widespread geographical coverage. If the reps don't produce orders, you pay nothing, although you will have training and management costs.

It is costly to train reps, so be sure to plan this activity carefully. After the training, you will find it necessary to establish an internal sales organization to support and manage the rep force. This additional expense of a few percent of sales can cause the cost of reps to be much higher than that of a direct sales force.

Although sales representatives do a good job of servicing existing accounts, they tend to be poor at prospecting for new accounts. It is therefore imperative to have a strong promotional program generating leads for them.

A sales representative firm will not always be loyal to your interests. After all, they have other principals' interests to consider, and there will be conflicts. At times, the self-interest of a rep firm may conflict with yours. Some conflicts will be in the form of cultural differences. Be sure that the rep firms you select have business philosophies similar to yours and that they project your type of image.

Product and market characteristics indicating the use of sales representatives are:

▼ If your product is easy to understand, then all the salespeople of the rep firms can be expected to push it well. They will also require less training.

▼ If your customers are numerous and spread over a wide geographic area, then the coverage provided by rep firms is attractive.

▼ If there are rep firms carrying products which complement yours in your market, then using those firms can be a powerful advantage for you.

So what's the bottom line? It's easy to see that most capital equipment businesses should have direct sales forces for greater control of your destiny, better customer contact, and lower cost. But for many start-up companies sales representatives are needed to reduce cash expenditures and lower risk. You should examine your business plan to see if the above indicators lead you clearly in one direction. Then, if direct sales is indicated, see if you can afford the initial costs of establishing the direct organization.

Selecting the Best Firm in Your Distribution Channel

Once you have selected the form of distribution channel that is best for your company, you are faced with the task of selecting specific companies. Here are some criteria to consider in evaluating a potential distribution partner:

▼ Do they have knowledge of your applications? Do they know a number of your potential customers? Most significantly, do they know much about your competition?

▼ Do they handle other product lines which are synergistic with yours? Any competitive lines? (Watch out!)

▼ How experienced are they? Do the key people demonstrate well-honed sales skills?

▼ How motivated are they to sell your product? A good indicator is their willingness to prospect for new customers. Are they willing to commit funds to promotional programs?

▼ How well does this company present the image you wish to create?

Relations with Your Distribution Partners

The relationship between your company and your distribution partners must be carefully forged to meet your needs. Most importantly, the relationship must allow you to have frequent and clear communication with the customer. Your distribution channel must not isolate you from the customer, or you might fail to perceive changes in the market or in your performance which could be detrimental to your future.

Communication with the distribution partner is critically important. Your people should accompany salespeople on selected customer calls, solicit their inputs on improvements to your company, and recognize outstanding sales performance when it occurs. You should have regularly scheduled seminars where sales people have the opportunity to learn of your future plans and even have some input into those plans. Treat them like partners in your business, and they will respond.

It is essential to train the people in your distribution channels well, not only for new products but old ones as well. Training seminars not only impart knowledge, but also communicate that your partners are important to you. These and other support activities raise your "share of mind" among the salespeople, usually at the expense of other product lines they carry.

You will need a sales support organization to back up your sales channel, even when your sales channel is direct. Salespeople invariably have questions on delivery, pricing, and technical issues. They also need coordination of large orders, special orders, and orders with different locations of specification, order placement, shipment, or billing. Summation, a Seattle-based manufacturer of modular test systems, recognized the need for strong technical support for their network of sales representative firms. Their solution was to place their own applications engineers in the offices of the reps to insure that immediate and clear support would be available when the reps needed it. The arrangement pleased everyone, especially the customers.

Probably the best received form of sales support is to provide qualified sales leads to your distribution channel. All sales organizations need leads, and they will use yours provided that you qualify them in some way. Most organizations will be willing to share some of your promotional expenses, and some will have mailing lists that you can use for joint promotion. You can make a big hit with your distribution partners by supplying them demonstration units, sales literature, and other helpful sales tools like custom-made demonstration boxes.

No matter what your distribution channel, you should carefully plan your administrative contact with the customer. When a customer places an order on your channel, it is important that the customer's credit status is decided quickly between you and the distribution partner. The processing of an order is an area in which you cannot afford mistakes. They can ripple through your entire organization and reflect back to the customer in the form of missed deadlines, wrong product configurations, and incorrect prices. After the shipment is made, you will need to coordinate accounts receivable with your distribution partner.

All distribution channels require careful management. You must agree upon sales quotas covering specific periods of time, and you should reset quotas when conditions change. Your distribution partners should supply you with forecasts of upcoming business, especially extraordinary events like large orders. Likewise, you should promptly report orders and shipments to them. There will be times when relations between you are strained. Be sure to jump on these issues and resolve them. Be businesslike in the negotiations, and don't be too quick to change partners.

Even though you have a sizable professional organization handling your sales, remember that the best salesperson for any company is the president of the company. You should be out in front of customers regularly, waving the flag and telling them how important they are to you. Good customer relations is part of everyone's job. All employees, right down to the finance clerk, should recognize their sales role. Telephone manners, quality of products shipped, correspondence. . . . all these contribute to happy customers. Try to create a truly customer-oriented company. It pays off!

Sell Around the World in Milliseconds

Our focus has been on developing a domestic sales organization. But the marketplace for most products is global. It only takes 130 milliseconds for a radio wave to travel around the world. Modern communications technology shrunk the world in the 1980s and made financial and business markets global. The United States is no longer the source of virtually all new product developments; we compete in worldwide markets and so should you.

1989 saw the start of many political changes that now make the development of foreign business more important than at any previous time. First, the European Economic Community started to eliminate internal trade barriers, making it the largest market in the world in the 1990s—over 320,000,000 people and $4,000,000,000,000 in purchasing power. Common telecommunications standards and approval processes, common import regulations, and easier movement of goods throughout the European community will characterize the new market. Companies that fail to establish a presence in Europe will miss a significant opportunity.

The rapid unraveling of the eastern European socialist economies presents another exporting opportunity for high-tech companies. These countries will desperately be trying to modernize their economies throughout the 1990s and will eventually be large consumers of high-tech products. A European presence will provide you with the opportunity to participate in the economic rebirth of eastern Europe.

The Far East also presents many export opportunities and competitive challenges. Japan, Korea, Hong Kong, Singapore, and Taiwan are countries that are aggressively competitive in high-tech markets, but that should not dissuade you from marketing your products in these countries. There are many products that they do not produce.

The lesser developed countries can also be good markets for high-tech products, but be prepared to work hard for your orders. The infrastructure is not as well developed and politics can play an important role in any buying decision.

Closer to home, the United States and Canada, its largest trading partner, signed the Free Trade Agreement (FTA) in January of 1989. The agreement established the following rules to open the borders for less restrictive trade:

▼ Eliminate most bilateral tariffs

▼ Establish "rules of origin" of manufactured products

▼ Eliminate import and export quotas

▼ Prohibit the use of non-tariff barriers in product testing

Prior to the treaty, there was significant preferential treatment for Canadian firms which included foreign content manufacture and pricing preferences.

In addition to increased sales opportunities, competing in foreign markets can help you stay in touch with developments before new competitive products appear on your door step. International sales will also develop a degree of economic diversification. Although national economies are loosely coupled, there are still timing differences that can help your company weather an economic downturn in the U.S. market. For example, Digital Equipment Corporation's sales were supported by strong foreign activity in 1988 when the U.S. market was extremely slow. Income from foreign activities was up 27%, while income from domestic operations was down 32%.

The economic reasons for international sales are very compelling. The export market for U.S. merchandise in 1987 was $253,000,000,000, of which 79% was high-tech machinery and transport equipment. "The President's 1988 State of Small Business" report to the Congress estimated that the total U.S. international trade will be 35% of Gross National Product by the year 2000, compared to 26% in 1987 and 12% in 1960. The Department of Commerce listed the following 15 industries as having the highest export potential; ten have significant high-tech content:

1. Computers and peripherals

2. Telecommunications equipment and services

3. Computer software and services

4. Medical instruments, equipment and supplies

5. Electronic parts

6. Analytical and scientific laboratory equipment

7. Industrial process control instrumentation

8. Aircraft, parts, avionics, and ground support equipment

9. Automotive parts, service equipment, and accessories

10. Electronic production and test equipment

11. Electronic power generation, distributions systems, and transmission equipment

12. Food processing, packaging equipment, and machinery

13. Safety and security equipment

14. Printing and graphics arts equipment

15. Water resources equipment

Foreign Sales Plan

How can your company effectively sell its products in the global market? Steve Weiner, director for the Center of International Trade Development in San Jose, California, says: "Have a vision to compete in a global marketplace at the beginning of the business, particularly during the product development cycle. Your funding sources should also be committed to the strategy." Too many companies do not consider the requirements for foreign markets during the development phase and try to sell their products to foreign customers as an afterthought. Consequently, many features that could make the product more appealing to foreign customers are overlooked or become too expensive to incorporate after the product is developed.

Avoid these expensive oversights and develop a plan to implement an international sales strategy at the start-up. The first step is to understand the international requirements for your product before development commences. A limited marketing study should be done for the product, at least defining safety and power requirements that may be unique in the foreign market. After the product requirements are defined, work can begin on the financial, product support, and sales details that must be addressed to develop a successful international business.

Bill Stolz, in his book "Start-up," describes how RF Communications, the company he started in the 1950s and grew to over 1500 employees, made exporting a cornerstone of its sales strategy. The company's second order was from Pakistan for $50,000! In some years, over 80% of the company's sales were to the export market.

DSP Group, a Silicon Valley designer of specialized digital signal processing chips, realized that large volume markets for their products would be in the Japanese consumer electronics industry. Davidi Gilo, the president, wooed Japanese customers when his company had annual sales of less than $4,000,000 in 1989. He had to visit some potential customers over 30 times, but it finally paid off. In 1991, his company was projected to do over $18,000,000, with over 70% of the business coming from Japan. Key customers included Kenwood, Sanyo, and Sony. In addition, Japanese investors provided over $9,000,000 in funding to the company. Clearly, it paid for the DSP Group to think internationally.

The development of an international sales presence is a long-term financial and emotional commitment. You must learn how business is done in a foreign culture and avoid committing cultural errors that can jeopardize a deal. Foreign businessmen have a long-range commitment to their markets and are willing to persevere to establish their presence for the long-term. You should too.

International selling is more expensive than domestic selling. Overseas airfares and higher expenses in major international cities contribute to high costs to visit prospective customers, attend trade shows, and meet with sales representatives. An investment plan for international business is just as important as a research and development or domestic sales plan. The goal is to build the customer's confidence in your ability to be a long-term, reliable supplier.

Post-sales support is another important area to be addressed in international sales. Inadequate support is often a major stumbling block for many high-tech companies; it can be fatal in the international market. A support plan that addresses the following items should be developed before a sale is made in any country:

▼ Local language manuals

▼ Spare parts

▼ Training of local technicians or users

▼ Factory support

If the sale is for a single product, the support plan could be to pro-vide 100% spares. High-volume products can support a local repair strategy which will eliminate many delays associated with a product failure. The following example illustrates short-sightedness on the part of the exporter. Handar supplied remote data collection platforms using telephone telemetry and battery power supplies to Panama. The customer was continually losing data that should have been collected by the system. After many weeks of long distance telephone calls, faxes, and efforts to reproduce the problem at the factory, Handar discovered that the customer had not properly configured the power supply system. Better documentation, training, or an application note might have averted this problem.

In addition to product support, you will have to promote your product in the foreign market. Product promotion will depend on the type of sales organization you develop. Distributors or joint venture operations will often develop local language brochures at their expense if they are convinced of the need for your product. Foreign representatives or direct sales approaches may require you to develop local language brochures. Important trade shows should be attended by you and your international staff to learn about the market and to support your local representatives.

A strong local sales presence will be the most important ingredient to initial success in the international market; effective management of the organization will assure its long-term success. The international sales effort can be managed by a separate sales manager, yourself, or be part of the domestic sales organization. Be aware that managing an international sales organization requires a high energy level and lots of enthusiasm. The manager must travel extensively and do business during business hours in foreign time zones from the home office. Whatever sales management choice is made, recognize that doing international business is different than domestic business and requires a greater commitment by your organization to the process. The credit decisions, shipping and delivery requirements, product configuration, and pricing require extra effort by your people. The effort to establish an effective international sales organization can fail without a commitment from your company's top management.

Paperwork is another very important part of international business. Payment for the products often involves a letter of credit drawn on an international bank. A letter of credit has important shipping and delivery conditions that control the payment for the product. Payment should be made in dollars with an irrevocable letter of credit (preferably drawn on a U.S. bank); as the business grows, foreign exchange may become important, and currency exchange rate fluctuations must be taken into consideration. Distributors normally handle all local collections. Be cautious in dealing with a distributor until a good credit history has been established.

Customs and shipping issues can be a serious stumbling block in the payment process. Select a good freight forwarder who can provide the trade documents at a minimum cost. The Department of Commerce holds classes to train your people on export documentation and export controls for your products. This information should be thoroughly studied as part of the process to develop international business. A lot of time could be wasted if a product that is intended to be exported is on a restricted list. This rapidly changing area should be constantly monitored.

If you anticipate a significant volume of international sales, you may want to establish a foreign sales corporation (FSC) which can provide certain tax benefits. This should be discussed with your attorney and accountant to determine if the benefits will justify the cost to establish and administer the corporation.

Sales Organizations

The type of representation you choose will differ between countries. Be aware of the alternatives and thoroughly investigate candidates by talking to prospective customers and other principals before any agreement is reached. In some countries, a representative agreement can be very difficult to break—it's a marriage for a lifetime.

The two most prevalent forms of representation for a start-up high-tech company are the foreign sales representative and the foreign distributor. The sales representative promotes products from different manufacturers and receives a commission on the products sold. The representative usually isn't responsible for collecting receivables or for verifying the credit of the customer. Choose a representative who sells complimentary products from other companies. The representative will be much more effective if your product can be sold to an existing customer base. The salespeople should be technically competent and should understand your product and its application, otherwise, you may have to provide continuous technical support to them halfway around the world. International customers expect a high level of technical competence from their sales representatives. This is particularly critical for high-tech companies.

Foreign distributors are similar to representatives, but usually buy and resell products with a markup to cover overhead and profit. They have the collection responsibility from the foreign customer. The distributor may stock products or buy only when an order is received from a customer. Be sure to keep track of the distributor's markup or your products may be priced out of the market. Most distributors will provide a level of after sales support and product promotion. Negotiate these items as part of the agreement with the distributor. Home office management will still have to provide training to the distributor's staff, attend trade shows, visit customers, and assist in the development of product promotion plans.

A direct sales force that works exclusively for your company can also be established, but this approach is very expensive and should be contemplated only if the sales volume can justify the cost. You incur all of the costs of promotion and trade shows associated with a sales representative plus the fixed cost of maintaining a full time field office in a foreign country. The additional costs can be over $150,000 per field sales person. Direct sales forces are usually established after a sales history has been developed using representatives or distributors.

There are also representative groups that are based in the United States and sell an array of products in various foreign countries. Some of these organizations can successfully promote your products through their extensive foreign connections. Dealing with a domestic foreign sales organization has some advantages.

▼ You're dealing with an American company

▼ They handle all export documentation

▼ There aren't language or cultural barriers

▼ It's like a domestic sale

There are also some disadvantages:

▼ They use agents in the country raising your price

▼ They are an additional middleman in the transaction

▼ Your company is more isolated from the end customer

Thoroughly investigate their background before making any commitments. Remember, your reputation will be established by a foreign customer's first experience with your company.

Another alternative is a joint manufacturing venture with an established company in the same or related field. You provide the technology and the manufacturing expertise and the partner provides the manufacturing facilities, product promotion, service, investment, and distribution network in the foreign country. Be sure to understand the capabilities of your partner to determine if you will benefit from the arrangement.

Handar established a joint venture with a Canadian firm to assemble, test, and market a sophisticated automated airport weather system in Canada. The attraction for this deal was a partnership with a local firm that was actively manufacturing and marketing aviation products in Canada. However, the sales advantage that was gained did not justify the amount of training, loss of marketing control, and support that was necessary to get the company up to speed.

Be sure to protect yourself so that if a joint venture deal fails at any time you will not sustain any losses, particularly when dealing with developing nations. Don't inadvertently create another competitor by protecting your worldwide marketing interests in the agreement.

The least complex marketing approach to achieve foreign sales is through a manufacturing license agreement. In this arrangement, a royalty is received on technology licensed to a foreign company. The negotiation of the licensing agreement and the technology transfer can be quite involved, but there will be less ongoing involvement in the sales and marketing of the products after the technology transfer than might be incurred in a joint venture. This approach can also yield benefits for your domestic operations with a cash infusion and through improved manufacturing expertise.

There is one other avenue to obtain foreign sales. You will often be contacted by people who claim to have an inside track on a prospective large foreign contract for your products. It is quite common to be contacted by three or four agents for pricing information on products when a large foreign contract is tendered. These people are usually opportunistic individuals with very little commitment to their customer or your product. Don't do business with them. Even if a contract is obtained, the burden of post sales support will be yours, because in most cases the agent that brought in the business will not be around to service the account.

Regardless of your choice of international sales representation, there is one constant: training. Whether you have direct sales people or foreign representatives, the success of your efforts will be proportional to the sales staff's understanding of your product and the competitive environment. This is especially true for high-tech companies. You must develop and implement a solid product training program and your representatives must be enthusiastic participants. The amount of training, the cost sharing, and the time commitment for both parties should be spelled out in the course of the negotiation and be documented as part of the sales agreement. Sales representatives who are unwilling to participate in a product training program are unlikely to be successful in selling your product.

Product pricing for international sales is another important part of a viable international sales business. Most high-tech companies charge 15% to 20% higher international prices to cover the additional marketing, commission, and service expenses. Foreign customers will occasionally try to purchase products for international delivery at a domestic address to get the lower domestic price. You should establish some internal controls to determine if a product sold to a domestic customer is really intended for international delivery. This precaution will protect your relationship with your foreign representative. If your product shows up at his customer, he may request payment of a commission if he has an exclusive sales agreement.

When selecting a foreign sales representative, remember that they will be making a statement about your company. The salesperson is the front line representative of your company. Keep that thought in the back of your mind during interviews with potential representatives. Customs may vary between countries, but be sure that a prospective representative is going to do business in an ethical manner. Develop a solid understanding of the type of working relationship your foreign representatives expect through personal dialogue and a well written agreement between the parties. Obtain and check personal and credit references of prospective representatives before entering into an agreement. Be sure to consult legal counsel to develop or review any agreement.

Worldwide markets can be pared down to size by assessing the areas that you are likely to have the highest success and establish your company there. For example, the majority of foreign manufactured laser diodes are produced in Japan. If your company makes laser diode testers, you should sell in Japan. Determine the best sales alternative and focus your efforts on establishing your position in the most promising market before going worldwide.

Getting Help

There are many sources of help for the development of international trade, such as the U.S. Department of Commerce, U.S. Small Business Administration, state export finance programs, local trade organizations, and even your local library.

The U.S. Department of Commerce, which has offices in 48 U.S. cities and 120 cities abroad, offers a variety of services for the prospective exporter. Market research information, including trade statistics, lists of competitors, trade barriers, distribution channels, and promotion practices is available through their International Marketing Research Center. This agency can also help locate customers and provide assistance in participating in officially recognized foreign trade events. The department conducts schools to train your staff on export regulations and documentation. Contact the district office in your area.

The U.S. Small Business Administration (SBA) is also a source of help and information on exporting. Financial help is available through the Export Revolving Line of Credit (ERLC) program in the form of bank guarantees for loans to finance labor and materials needed for manufacturing for export and to penetrate or develop foreign markets. Contact the local office of the SBA for a list of services available in your area.

Some states, such as California, have established export finance programs to augment the financing efforts of the federal government's Export-Import (EXIM) Bank. For example, the California Export Finance Program is intended to make it easier for small businesses to increase their export sales through financial and information assistance. The program provides loan guarantees and counseling to help manufacturers utilize other state and federal programs for export. Contact the SBA or Department of Commerce office in your area for program information.

A few communities have begun to take a pro-active approach to stimulating foreign business for local companies. San Jose, California has created the Center for International Trade Development (CITD) with three goals:

▼   Help local companies learn how to do international business

▼ Develop and maintain strategic relationships with international cities

▼ Stimulate overseas investment in San Jose

The relationships that are developed at the political level can help to quickly open doors for a new company wishing to export products to the foreign country. For example, as a result of a trade mission trip by the mayor of San Jose to Osaka, Japan, a local software company was able to conclude negotiations in one night on a strategic alliance with a Japanese company that they had been working on for over 17 months.

In summary, the development of an effective international sales organization can be a challenging job, but it can be both personally and financially very rewarding. Finding new markets for existing products is often a much easier task than developing new products to provide the same growth and profitability, particularly if attractive features for the international market are incorporated into the product in the development phase. Appendix E provides an outline for a plan to proceed. Remember, in the early stages of the company you are using the leverage of existing research and development, domestic marketing, and manufacturing to increase revenue through international sales. That leverage can lead to excellent bottom line results.

Foreign customers can be more demanding than domestic customers with regard to product features. This can mean a requirement for more special modifications to your products to satisfy international requirements. The next section will help you plan for both international and domestic specials.

Some Products Are Really Special

What do you do if a customer wants you to modify your product just for them? All high-tech manufacturers eventually have to face this question. Obviously, you can turn down the business. But it's not as simple as just saying "no." In some cases, special modifications will be an important part of a business strategy. It is essential to identify the extent to which your customers will require special modifications. Then you can decide how to deal with the issue in your company.

There are businesses in which modifications are the rule, not the exception. Some equipment for scientific research or industrial measurement is customized for every order. OEM businesses often modify products for each major contract. Many industry pundits, including Tom Peters, believe that this is how most products will be sold in the 1990s and beyond. Businesses like this need to be structured flexibly to respond to the unique demands of customized orders:

▼ Manufacturing must assemble products by order, rather than scheduling "runs" of identical finished products to be placed into inventory.

▼ Your product development department will spend much of its effort designing accessories which can be used to customize standard products. These accessories sometimes become big sellers.

▼ Your order processing procedures must be designed to produce complex quotations, make requirements clear to manufacturing, and check for errors. You will need to apply resources to accomplish this.

▼ Accounting procedures can be quite different when your products are made-to-order. You will need project cost tracking and finely tuned overhead allocation, among other refinements.

What about a business that has only a minor content of specials? Some organization will be needed here as well:

▼ Documentation of specifications, operating procedures, and repair procedures. Ask yourself how much documentation will be required. Who will do it? Will it be part of the original user's manual or a separate package? Be sure to consider the cost of documentation when preparing quotations.

▼ Assignment of a responsible party to design and perform the modification. This doesn't have to be the same person each time. Be sure that the necessary resources are available for the job.

▼ Agreement by all departments that each modification is acceptable and that the necessary resources will be committed. Don't forget manufacturing!

▼ A label in the manual and on the product, so that service technicians in the field will know what they're dealing with.

▼ A plan to support each modified product, written before the product is shipped. It should cover topics such as extraordinary installation procedures and customer training.

No matter what your organization, leave product development up to the product development department! The issue here is quality as much as allocation of resources. Product development projects include extensive testing, some under severe environmental conditions, to insure product reliability. You usually can't do that for a single order. So, if you perform special modifications for customers, they must have only slight impact on the product's performance. Also, they must be simple enough that only nominal engineering time is required. Remember, this engineering effort will result in only one order, whereas new product engineering has leverage in the sales of many units.

Sometimes, when a special modification has significant market potential, you may convert it to a standard product. Any project to do this should be conducted by the product development department, again to insure quality. These product "enhancements" can be beneficial, but you must be on guard for the fallacy of the incremental investment. It is easy to convince yourself that a small engineering investment will yield a large return. But the fact is that each succeeding small investment has a smaller return than you think, since preceding changes may have already appealed to the same customer. The diversion of engineering resources toward an enhancement carries the very large cost of new product delays. It is often best to pass up enhancements of present products in favor of completing new, higher return products sooner.

Specials can be a good source of revenue, product enhancements, and warm customer relations. But you have to handle them correctly, or they can drain your profits and swamp your company with problems later on.

Product Support Sells, and Sells, and ....

Product support, particularly for specials, is often one of the last areas to be addressed in the evolution of a high-tech company, but it can be a source of significant competitive advantage. Consider your reaction when you purchase a product that requires some familiarity or experience to operate and the owner's manual is incomplete or unclear. Most high-tech products are far more complex to use than consumer products, but many companies give only passing consideration to this facet of their business.

For example, we know of a $30,000,000 public company that routinely spent over 17% of revenues on product development but often did not have manuals completed when products were ready for delivery to customers. In fact, when questioned about the priority for customer documentation, the answer was that the manuals generated by the engineering department were sufficient. Product support had low priority. Consequently, salespeople were often required to do extensive customer support rather than sell products, and engineers had to provide applications advice rather than develop new products.

The development of a strong product support group is a good investment for your company. Immediate returns can be achieved in the following areas:

▼ *Repeat sales.* Satisfied customers are more likely to make additional purchases.

▼ *Relieve sales personnel from post-sales support.* Competent product support personnel can relieve your sales people from the task of helping customers get up and running with a new purchase.

▼ *Competent understanding.* Product support people should be very competent in the operation and use of your products. They will develop experience with the kinds of questions that most often occur regarding its usage and can provide a high level of comfort for your customers that will facilitate future sales.

▼ *Customer contact.* The product support group will often have more customer contact than your sales people who are often traveling. The frequent contact will provide feedback on the quality of manufacturing and design, and on future product enhancements.

A close working relationship should exist between product support, manufacturing, engineering, and sales. Problems with products in the field that relate to design or manufacturing deficiencies should be dealt with in an atmosphere of teamwork. Product support should not be a scapegoat for shortcomings in other departments. It should exist to enhance your customer's use of a well designed and well manufactured product and not be an antidote for poor product quality.

For example, much of Handar's success could be attributed to the development of a positive attitude toward product support. Many of Handar's customers were large U.S. government agencies that influenced foreign sales and other new agency purchases. Product support personnel often spent time in the field helping customers deploy equipment. As a result of this extra effort and personal interaction, Handar learned a lot about its customers' field problems, and the customers felt comfortable recommending the equipment to other agencies.

Customer Training

Training courses help customers maximize their use of your products and are a good sales tool as well. Customers will feel more comfortable with your company because of additional personal contact and will appreciate the extra knowledge that can make them more productive. Customer training also provides feedback on product applications, the design quality of the products, and future product requirements. The exposure to other products in the course of training can stimulate additional sales.

Another benefit is a reduction in the amount of post-sales support for a product. TSI, Inc., a St. Paul, Minnesota manufacturer of fluid flow instrumentation, developed a training seminar that was typically held at universities to train researchers on the latest commercially available technology. The setup of a laser doppler velocimeter, a $150,000 device used to measure turbulence in fluid flow experiments, was so complex that graduate engineers working on their Ph.D. degrees needed training from the company. The training program significantly reduced the amount of on-site support required by the customer and provided an opportunity for company personnel to interact with researchers that were actively involved in state-of-the-art measurement problems; what's even better, the customers paid to take the course!

Customers will pay for value and training has value. Training at most Fortune 500 companies is viewed as a profit center; your company should do the same. If your industry has set a standard for free training, consider including the cost in the pricing of the product.

Documentation

The product support group can contribute significantly to a product's success through the quality of the user's manual. Many high-tech start-ups make the mistake of having engineering personnel create user's manuals. They are the last people to consider for this task; they are too familiar with the product to develop a good manual. The receptionist could probably write a better manual. Engineering should review the content and should edit the final draft, but a good manual writer should develop it. If you don't have a person like this on your staff, hire a consultant. It may seem like an unnecessary expense at the early stage of the company, but you will reap the rewards later in more satisfied customers.

Engineering change notices (ECNs) are related to the manual development process, but are often viewed as an internal affair. A manual for most products should be viewed as a living document and a procedure should be developed to keep it current with the products being delivered to your customer. Large organizations, such as the U.S. Defense Department, that buy large quantities of products may insist on configuration control of the hardware, software, and manual. This requirement can place stringent control on the timing and method of implementation of any changes to a product or its documentation and must be carefully managed. The product support group should be part of the ECN reporting process so that new customer documentation can be developed as changes are made to the product.

Software bugs have become a significant factor in product support. Many electronic products contain a microprocessor and software. Software is not only difficult to design, it is difficult to document, especially if the product can be custom-programmed by the customer. You should establish a "bug-fixing" group if you have adequate resources or assign the task to development engineers and take the time spent fixing bugs into account when planning new project schedules.

New Product Development

The product support group should have a representative on all new product development teams. Product support engineers are on the front line in the trenches dealing with all of the inconveniences and shortcomings of your products. Customers seldom call extolling a product's virtues. Take advantage of this experience and give the product support engineers a strong voice in the design of your products. You will be pleasantly surprised at the number of practical ideas you will receive that can significantly reduce after sales support cost, reduce repair costs, and improve the product's function.

An example of the progress in product support involvement in new product development is the Xerox announcement in early 1990 that their new copiers could call the service department before the copier fails to meet customer expectations. This kind of internal diagnostic capability can significantly reduce customer service costs and enhance customer satisfaction, often with little additional manufacturing cost.

Product support is an important component of your company's structure. It can pay big dividends in customer loyalty and repeat business, but it need not be costly to implement. Consider selling services wherever practical or including the cost in the product pricing strategy. Customers will usually pay for support directly or in the purchase price if the total experience with your products is positive.

CHAPTER 7

Growth Through New Product Development

You have written your business plan and convinced investors to fund your new venture; you're on your way to running your own show—or are you? Many high-tech electronics start-ups get to this point and then fail to execute the plan that they sold to their investors. They don't meet their commitments to develop the product they described with the allocated resources in the agreed-upon time frame. This failure usually leads to—at worst—shutting down of the company for lack of funds or—at best—significant dilution of the founder's equity in the company because of needed additional financing, often on terms unfavorable to the founders. How can you avoid this trap?

First, you should be as realistic as possible in estimating the amount of time and money required to develop the product. Engineers are notoriously over-optimistic when estimating the time and, consequently, resources required to complete a project. You have to protect yourself by putting in financial and time-line padding in your plan.

Many projects have a critical technology path that can significantly delay the project completion. If there is an area of significant risk or technical difficulty, try to have these technical questions answered before you staff up the whole project and other areas of the company such as marketing or manufacturing. Getting answers in advance can significantly reduce the "burn rate" of your capital.

Take into consideration the significant factors that can influence your ability to get the product to a shippable condition. Does it need Federal Communications Commission or Underwriters' Laboratories approval? Are you planning on selling initially in Europe or other foreign markets? If so, have you considered the impact of approval by agencies in those markets (such as the post office, which must certify

143

or approve telecommunications products in many countries)? If you don't have answers to these types of questions, or worse, you don't know which questions to ask, you should do some further planning. Your investors will not necessarily raise the appropriate questions.

Many engineers think that by leaving the larger corporate world and doing their own thing they can leave the discipline of schedules and budgets behind. Nothing could be further from the truth. In a start-up you have a finite amount of money to achieve your objective in a defined time. If schedules slip, you'll need more money which will come at a dear price. Missed deadlines can make you miss a market window and sink the whole enterprise. Get your development team to buy into the schedule and make sure that they understand the ramifications of delayed product introductions. A product delay can significantly impact the market launch and slow the ramp up of sales.

Cash should be conserved in a start-up. Make sure that your team has the necessary tools to do the job properly, but don't get carried away with too many frills. Certain items, such as efficient software development and computer aided design tools, are essential to get the job done, but consider the use of outside services if at all practical. Consider leasing or renting equipment to conserve cash when equipping your team.

Software is a significant development bottleneck for most products. The product features, including the user interface, should be clearly defined before any design proceeds so that the developers have a clear understanding of how the product will function from the user's perspective. This can be achieved by writing the operator's manual before any code is written.

A development schedule should be produced that addresses each functional area of the software, and the staff should be managed to make sure that only those features that are in the original design get implemented. There is a tendency for products to develop "feature creep" as the development team comes up with new features that would be "neat" to have in the product. This temptation should be avoided if you want to meet your deadlines. It is usually possible to introduce upgrades or "B" models after the first product is on the market when you have some cash in-flow to fund additional development.

Unexpected problems can also derail an otherwise well-planned development program. Try to anticipate some of the areas that might cause trouble and make a contingency plan to avoid being blindsided by a glitch in the program.

New Product Development Is Your Fountain of Youth

After you have launched your first product, you will have to do an encore. A product development program should provide a stream of new innovative products for your company. This point seems very obvious in the start-up phase of a company, but keep in mind the story of Control Data Corporation (CDC), one of the original high-tech electronics stars which was a start-up in the 1950s. CDC had a significant share of the large mainframe computer market; its CYBER computers were the machine of choice for large number-crunching applications such as weather modeling and geophysical data analysis. The company started to lose its way in the late 1960s when it began to diversify into nonrelated businesses such as credit union services and financial products. CDC did not keep up with technology and lost market share to Cray Research, a supercomputer spinoff from CDC founded by Seymour Cray. The company also lost market share to the new technologies of networked personal computers and workstations. As the 1990s began, CDC was struggling for survival primarily due to a shortage of successful new innovative products in its core business area.

In contrast, the growth of Tandem Computer, the world leader in redundant computing by 1990, has been a highly successful model of high-tech entrepreneurial success. The company's stated strategy in its 1974 business plan was: ". . . to focus corporate resources totally on multiprocessing, so that at all times Tandem will have relatively larger development programs and marketing efforts aimed at this sector than any other competitor." The business plan described how the company would continue to invest in an array of innovative new hardware and software products to secure a beachhead in the highly competitive computer business.

Most successful new companies are started with products that are significantly different from current offerings. This is accomplished with a good product definition that satisfies unrealized customer needs. The product should have significant technical, quality, cost, and feature advantages over competitive products. If not, you risk rapid retaliation by the competition. If your product depends on special technology, be sure that the technology is not about to become obsolete or be bypassed by advances in standard technology that you have not anticipated.

A good example of a technology that looked like a winner in the late 1970s was bubble memories. Hewlett-Packard and Intel both spent a lot of research and development money on the technology. Unfortunately, the cost/performance of rigid disk drives improved much more rapidly than anyone anticipated and eliminated any significant demand for bubble memories.

Undifferentiated products cannot be saved by a large advertising budget. Regis McKenna stated in his book, *The Regis Touch*: "With new technologies, differentiation must begin with the product. Companies must start by giving tangible evidence of the product benefits. Companies can't just go out and say: We are the leaders. Intangibles, such as leadership image, must grow out of the tangibles." This is an area that professional investors in venture capital firms will focus on. They will want to know what your sustainable "unfair" competitive advantage will be.

Few high-tech start-up companies have trouble focusing on the development of the first product; it's why the company was started anyway. But even very successful start-ups can lose their way. In August 1989, 3Com, one of the leading suppliers of computer networking products, announced its first quarterly loss after five years of sensational gains in revenues and earnings. Part of the problem was a general weakness in industry orders, but Bill Krause, their CEO, told the San Jose *Mercury News* that the company had not introduced enough attractive new products to help it weather the storm as it had in the past. Krause said, "Our biggest problem is no new products." As evidence, 3Com introduced 19 new products in the previous fiscal year, but had only four scheduled for the new year.

On a smaller scale, Handar continually invested 8% to 10% of sales on product development each year. However, the company lost its focus during one product development cycle and launched an ill-fated project to diversify into industrial data acquisition products. During this time, the company neglected product development in its core business in environmental data acquisition products. The diversification failed because of poor product definition and an inadequate sales channel. The lack of new products allowed several competitors to gain market share in the environmental market. After the mistake was recognized, the company got back on track and developed an automated airport weather station that included an infrared visibility sensor and a solid-state laser cloud height sensor. The company was subsequently successful in winning a multi-million dollar contract to supply the visibility sensor to the FAA for upgrading all runway visual range systems in the United States.

Even highly successful companies that invest heavily in new product development can face a problem of a different nature: intense competition. Sun Microsystems, the leading computer workstation manufacturer in 1989, had a reputation for introducing new products at a rapid rate but Digital Equipment Corporation, Hewlett-Packard, and IBM were close on its heels. Sun announced six new products in April, 1989, but within the next six months, H-P responded by acquiring Apollo, a workstation manufacturer, and DEC announced a new product that was $1000 less and 35% more powerful than Sun's product.

Trimble Navigation is a company that made research and development investment the cornerstone of its strategic plan to grow and maintain market share. The company spent more than 25% of sales, over $10,000,000 per year at 1989 revenue levels, on new product development. The company was driving the market growth at more than 30% per year with new, lower cost technology for satellite navigation and wanted to raise the ante for the competition to stay in the game. This high investment strategy drove the market from a military orientation to a large commercial market, including automobile navigation devices and vehicle location systems.

Develop a Strategy

The message in these stories is that you should have a product development strategy for the company and stick with it. You may develop the market with your initial product and not reap the rewards because the initiative is snatched away by a competitor that effectively commits more resources to product development than you do. This is not so obvious in the start-up phase because the whole company is focused on getting the first product developed. However, after the first product is completed, the question is "What do we do next?" You must have a plan both for new product development in your company and a general idea of the complimentary products to be developed after the first one is completed.

The level of investment required to develop new products varies widely between industry segments and even within the segments. In a 1986 survey by *Electronic Business* magazine of the top electronics spenders on research and development as a percentage of revenues, investment by the top 100 publicly held electronics companies ranged from 24.3% for GCA Corporation, a manufacturer of semiconductor processing equipment, to 3.6% for AST Research, a manufacturer of personal computers and personal computer add-on products. The range of investment within industries is also dramatic, as shown in Table 7-1. Keep in mind that this table was based upon a 1986 study, and the figures shown do not necessarily correspond to the current investment levels of the companies. (In fact, some of the companies—such as Apollo and Burroughs—have since merged with other companies.) However, these figures show there is no "right" level of investment for any company.

TABLE 7–1

Industry	High %	Company	Low %	Company
Semiconductors	17.6	AMD	5.3	General Inst.
Mainframes	15.6	Amdahl	5.7	Burroughs
Capital Equipment	24.3	GCA	7.5	Varian
Components	9.8	AMP	3.6	AVX
Minicomputers	14.1	Apollo	4.2	Telex
Microcomputers	7.1	Altos	3.8	Apple
Peripherals	9.7	Storage Technology	4.6	Tandon
Communications	20.1	DSC	5.1	M/A–COM
Office Equipment	9.0	CPT	4.0	Diebold

It is imperative that you have a research and development investment strategy that will provide you with new products to maintain a leadership role in your market niche. Remember the Tandem example earlier in this chapter? Tandem expected to invest more in product development in their niche market than their much larger competitors, which included Digital Equipment and IBM. Tandem was highly focused on its goal. Thus, even though the company was a start-up, it planned to have superior development resources than the competition in its niche.

 The amount of investment is not the only criterion for success. The money must be spent effectively and the company should be in a market that justifies the level of expense. Note in Table 7-1 that Altos Computer outspent Apple by nearly two to one on a percentage basis, but it could not leverage that higher investment into high growth in their market.

It is common to spend considerably more than industry averages on product development to establish a product line and gain market share in the early years of a company. As the company matures, the level of investment usually changes, but it must still be the cornerstone of your strategic plan.

Developing new product capability need not be constrained to your research and development lab. Acquisition of companies or technologies that fit your strategic plan is another viable alternative to generate growth.

Another method for increasing your research and development advantage is through technology licensing from university research work. Most major universities have active licensing programs to move technology from the laboratory to the marketplace. You can get information on such programs by contacting universities directly and by attending conferences where papers are presented by academic experts. This source of technology should not be overlooked; recall that Sun Microsystems was started using technology developed at Stanford University (namely, the Stanford University Network).

We've spent some time talking about developing new products through the research and development process. Remember in Chapter 6 how we discussed the new product introduction plan? Companies often lose their way when they have some success and the research and development department becomes flush with money and new ideas. This is the time when the "automatic cabbage picker" or equivalent product is conceived. Technical entrepreneurs often become enamored with their ability to create new products, and forget about how the products will be sold or marketed. They think "hey, we built the world's most successful personal computer, why not an automated cabbage picker?" They fail to recognize that the distribution channels and marketing efforts might be significantly different for their new product. Throughout the new product development process, you must align new products with your existing sales and marketing capability or be prepared to make massive investments in a new organization.

Gelber's mistake on ModelMaster — not understanding it was different from a power meter.

Duplicate the Success

The definition of the first product is usually not a problem. Everyone associated with the company will know what is being developed, for what customer, and many will have had a hand in defining the product, even if it's only the product's color. You will personally be very close to the technology and have intimate knowledge of the intended customer's requirements. How does this intimate understanding of the customer's requirements get transferred to the engineering department when you are no longer the chief engineer or chief product architect? How can a process be developed to assure that your company will be as responsive to customer needs when there are 1000 people as when there were 10?

First, you must have a commitment to developing manufacturable products that customers want. Second, and almost as important, your engineering vice president should share your commitment. If he or she doesn't, look for a new one. Your engineering vice president is one of the most critical people you will hire or have on your founding team. He or she should have a good strategic planning mind; otherwise, you and your marketing vice president will be faced with developing and justifying your product strategies to engineering. It is very difficult to get commitment to product development plans if the engineering vice president is not a major contributor to the plan.

Your engineering vice president should also be committed to product development as an element of a business, not just an outlet for creative talent. This means a commitment to working with other departments such as marketing, manufacturing, sales, and product support to facilitate the development of the best possible product. Products that can't be sold due to inadequate specifications or can't be manufactured due to poor design can quickly ruin your company.

Finally, the engineering manager should know how to manage engineers to stimulate creativity and get products out the door.

Some managers try to use a schedule as a whip to get a product released to manufacturing. This rarely works—engineers are motivated to complete projects. It takes management skill to combine the creative process with timely development of new products. Here are a number of suggestions to help you create this environment:

▼ *Investigation/Development Program*
As your research and development program matures, some portion of the budget should be devoted to exploring new technologies or product ideas that are related to your business but may not necessarily become products in the short-term. A small team or even a single person can make remarkable headway on solving problems in an investigation. For example, in 1975, an investigation by one person into microwave counting technology lead to a patent for a single sampler microwave counter with higher performance and lower cost than current products at Hewlett-Packard. The initial investigation was done at low cost because the staff level was kept to a minimum. The proven technology was then applied to several development programs which produced an array of new products for the company. The technology was still in use 15 years later in a variety of products. Development teams should also be kept small to minimize communications overhead.

The federal government's Small Business Innovative Research (SBIR) program is designed to encourage small businesses to develop technology that can be commercialized. You may be able to gain some development leverage by finding technology areas that are of interest to a government agency and can also enhance your capability. TSI, Inc. took advantage of these programs to further their technology base in the area of particle counters and laser based measurement systems. In order to participate in this program, you should find an agency that may have a need for your type of technology and propose a study or submit a proposal for a study the agency has defined. Initial contracts are awarded for up to $50,000, with follow-on contracts of up to $500,000. The Small Business Administration can provide more details on the process for securing an SBIR contract. However, do not get so embroiled in this type of contract work that it begins to negatively impact your regular commercial development.

▼ *Tools and Support Services*

Provide the necessary tools and support services to get the job done in research and development and in the rest of your company. This doesn't mean a lot of frills, but don't scrimp either. Particular attention should be given to providing the appropriate software development tools. Fast compilers, debuggers, editors, and workstations can greatly facilitate the development of product software.

▼ *Design Reviews*

The project team reviews each engineer's design to provide constructive feedback on the design and to help each engineer better understand how his or her work relates to the whole project. Establish an environment where design reviews are viewed as a learning experience, not as the Spanish Inquisition. This process can avoid costly mistakes and greatly improve the quality of product designs. In the process of explaining a design, an engineer can often learn much more about the design and the conscious or unconscious tradeoffs that were made. The process also improves team communication and understanding of the project. The design review is also an excellent way to facilitate teamwork.

▼ *Process Knowledge*

Your research and development engineers should understand the manufacturing technology that is employed in your company and stay current on new developments in the field. This can be facilitated by frequent interaction with manufacturing personnel and the inclusion of manufacturing representatives on the development team at the earliest possible time. In the early days of Hewlett-Packard, most engineers spent their first six to nine months of employment working on a factory assembly line. This experience gave the young engineer a perspective of what it was like to build hundreds or even thousands of the same product. In today's competitive market, the product not only has to be well designed with the right features, it has to be easily manufactured.

▼ *Product Support*
Product support has its own process that may impact design decisions. The repair and support philosophy for a product should be considered at an early stage of product development. For example, a remote diagnostic capability may be necessary to provide adequate response time to field failures. This strategy would fail if no communications capability were designed into the product. Product support personnel should also be represented on the development team at the earliest possible time. The engineering staff should be committed to supporting manufacturing, product support, and marketing throughout the life cycle of the product.

▼ *Continual Change and Improvement*
Americans have become accustomed to the "big bang" approach to new product development, such as the transistor, the integrated circuit, and the microprocessor. While these revolutions will still occur, the more common developments are characterized by continuous evolution of products and technology. The Japanese have been the most consistent practitioners of the process in the 1980s. Engineers look at most new products and say: "If we just had six more months. . . . " Your competitors will look at your products in the same way. You must have a plan to obsolete existing products with new improved products on a scheduled basis. If you don't, your competitors will. The schedule will depend on the rate at which your competitors introduce new products or how aggressive you feel you must be.

▼ *Customer Feedback*
Take every reasonable opportunity to expose your engineering staff to customers. Have them participate in customer training seminars, make customer sales calls (with salespeople, of course), and make service calls. Any interaction with customers improves their understanding of the needs of the people your company is trying to serve. This is how most successful companies start, but they often lose the way as they grow.

Corporate standards for colors, logo, and product appearance should be established when you develop your first product. A good industrial design firm can help immensely, but manage their efforts closely or you may be unpleasantly surprised when the bill for their services arrives. We know of a firm that had over $20,000,000 in revenues and nearly 100 products before any corporate standards were set. Each engineer got to choose the cabinet style, the color, and even how the logo appeared on each product. Needless to say, the company had very little product identity in the marketplace.

Making It Manufacturable

The products you design must be manufacturable at low cost and at the highest level of quality. Design for manufacturability and assembly (DFMA) is extremely important to gaining and maintaining a competitive edge in the market. At an American Electronics Association (AEA) symposium on productivity in 1988, it was reported that 85% of product cost is determined at the design stage. For example, a design with fewer components reduces assembly time and lowers direct product cost. Fewer direct labor workers are required which requires fewer overhead supervisors and clerks. The cost reduction flows throughout the company.

Lower part counts will also improve reliability. If a product has 400 components, each with a probability of being good of 0.999, then the probability of the product being good is $(.999)^{400} = 0.67$. If the number of parts is reduced by 25%, the probability of a good final product increases to 0.74, a 15% improvement.

Bill Schroeder, president of Conner Peripherals, reported on Conner's cost minimization results at the same AEA symposium. The company competes in the highly competitive disk drive industry and their achievements as outlined in Table 7-2 for the quarter ending December 31, 1987 set a new industry standard for financial performance through careful design and manufacturing techniques:

TABLE 7-2

Conner Peripherals' Results Compared to the Disk Drive Industry for the Quarter Ending December 31, 1987

	Conner	*Industry Composite*
Revenues in millions	$51	$132
Operating margin	17.2%	9.3%
Inventory turn	6.7	4.6
Fixed asset turn	12.3	6.0
Return on assets	110%	23%
Revenue per employee	$155,000	$87,000
Net profit per employee	$16,900	$6,400

Besides books and articles on DFMA, there is also a software package. Boothroyd and Dewhurst, a Rhode Island company, specializes in the development of design for manufacturing productivity software, which utilizes an extensive database of component, process, robotic assembly, and manual assembly costs in a rule-based system to analyze product design decisions in the early design phase. This technology provides designers an early look at the effect of their decisions on the manufacturing cost of a product long before the design is finalized. The results are significant. NCR utilized this technology in 1988 to develop a new point-of-sale terminal with the following results:

▼ 44% reduction in overall manufacturing labor cost

▼ 85% fewer parts

▼ 75% faster assembly time

▼ 100% reduction in assembly tools

▼ 65% fewer suppliers

These results show that product development is more than creating new widgets; the design process controls the cost structure of your products. The program should be guided by focussing on *innovation, quality,* and *productivity*. Your commitment to these ideals is imperative and should be articulated in a small manual in the start-up phase that outlines the process and defines responsibilities for new product development. The manual should be a living document and should be developed in the spirit that it will help each member of the development team and ultimately the company do a better job (Appendix G). By staying involved with the process and providing constant encouragement, you can instill commitment to the ideals.

CHAPTER 8

Setting Up a Manufacturing Operation

*P*roduct development usually has the limelight at most start-up companies. There is excitement in the air as the product starts to take shape. The question of how the product will get built is often not considered at this stage—development engineers often think that just automatically happens. This attitude will not empower your company to become a world-class producer. Manufacturing expertise must be brought in early and be part of the development process.

This philosophy runs counter to the attitudes found at all but a few of the best manufacturers in the United States (it's reflected in their salary structures), but consider the advantages that companies focussed on manufacturing have over their competition: lower production cost, higher quality products, higher morale, and lower financing costs. Manufacturing capability is often as much a part of your strategic advantage as the unique technology of your product. As a market matures, and they inevitably do, manufacturing capability and an ability to quickly introduce new products into manufacturing will become even more important to your company.

Conner Peripherals was mentioned in the last chapter as an example of a company that has made cost minimization through careful engineering practices one of their key strategic advantages. Manufacturing was also part of this strategy. The production and test process was central to the operation of the company, with tight feedback between production and design and an emphasis on continuous process improvement. The company focused on building a world class manufacturing organization to produce the highest quality and performance 3.5-inch disk drives at the lowest cost.

You should carefully define your business before attempting to set up the manufacturing organization. Understanding the type of business will influence many decisions for manufacturing. Is it going to be a custom systems business or will you build standard products? If you are producing standard products, are they shipped from stock or built to order? For example, Handar builds standard products that are assembled to order. The company has a catalog of standard data acquisition systems with signal conditioning, communication, and sensor options. When the company receives an order, the system is configured, tested, and shipped to the customer.

In contrast, Lightwave Electronics builds its products as custom sys-tems. Each order is tracked on a large board in the production area by serial number. Each product has its own manufacturing group that is responsible for building the product to meet delivery requirements. The technical experts that designed it are part of each group and share responsibility for meeting shipment deadlines.

Apple Computer builds standard products that are shipped from stock. Their Fremont, California manufacturing facility used the most advanced systems available in the electronics industry at the time it was opened in 1990. Vendors had to meet precise delivery times within 15 minutes, two or three times a day, to assure that parts were available for the assembly line when needed. This system, called just-in-time (JIT) inventory management, allowed Apple to minimize inventory at their plant. An inventory and production management system this aggressive can only be used for producing standard products in high volume.

In addition to being well organized to complete its primary mission to build quality products, manufacturing should have a customer focus. In many corporations, the manufacturing operation is isolated from the customer; the people building the products don't know who uses them, how they are used, or what happens if they fail. Manufacturing employees should understand that customers expect high quality, low cost, and on-time delivery.

Be aware that there are many conflicting objectives within an organization that will make the optimization of manufacturing plans very difficult. For example, consider the matrix of priorities for the various groups within a company as shown in Figure 8-1.

The top line in Figure 8-1 indicates the company's objectives and the bold entries in the table show where objectives of the groups in the organization may be in conflict with the company's objectives. For example, marketing wants to maximize customer service, which is in agreement with the company's goal, but this objective can drive up product cost through model proliferation, increase inventory because of excess stock on hand to meet all delivery requirements, and reduce factory utilization because of a multiplicity of products and options. Your management skill will be required to get the various factions of the organization to work together to optimize the organization's objectives, not just their department's objectives.

OBJECTIVE	PRODUCT COST	CUSTOMER SERVICE	INVENTORY	FACILITY UTILIZATION
COMPANY	Low	High	Low	High
MANUFACTURING **Long Runs**	Low	**Low**	**High**	High
MARKETING **Max Customer** **Service**	**High**	High	**High**	Low
FINANCE **Low Product Cost**	High	**Low**	**High**	High
FINANCE **Min Asset Cost**	**High**	Low	Low	**Low**

Figure 8-1: Developing manufacturing plans for a company requires managers to balance many competing forces.

We have stated that your company should strive to be a low-cost producer in its industry. This is often thought to be the exclusive domain of the manufacturing department. Certainly, much of it is, but consider the effect that decisions and actions by other departments have on production cost. Research and development can design a product that has inadequate documentation, inadequate design margins, or difficult assembly processes. Marketing can require too many options on a product that too few customers want, which reduces the volume on standard products. The most conspicuous example of this effect is in the "options proliferation" on cars produced in Detroit. The number of possible configurations on just one car model in 1989 exceeded 1,000,000 when paint, upholstery, and mechanical options were included!

Cost control should be on everyone's priority list, not just manufacturing. A study done by the American Production and Inventory Control Society (APICS) in 1987 showed that the top four contributors to excess cost were engineering change orders, product proliferation, option proliferation, and vendor proliferation. Manufacturing only has control over vendor proliferation. They do have control over many other production costs and should be organized to take advantage of the latest methods which include product line teams, manufacturing cells, and shared responsibility. Cost allocation should also be addressed to make sure that the impact that various products have on manufacturing cost is fully understood. Low volume products that appear to be profitable when riding on the coattails of high volume products are often found to consume a large amount of overhead when evaluated independently.

Be sure to keep your quest for low manufacturing cost in perspective. The important measure is the success of your product in the marketplace and the price the customer has to pay for it. The price has to take into consideration the cost to service, support, and market the product. So there are often many trade-offs to be made in attaining the lowest total cost. You may want to put more resources into design, service, or marketing rather than spend more on beefing up manufacturing to attain an additional 5% factory cost reduction.

After product introduction, you will be faced with figuring out how many units to build per month—you need a sales forecast. The sales forecast is translated into a master production schedule (MPS) which determines how many units of each product to build. You have now stepped onto the tightrope that never ends—reconciling sales forecasts to production schedules. This is where BIG money is made and lost in manufacturing businesses. Commitments will be made from sales forecasts to purchase inventory, equipment and space, and to hire personnel. The development of this plan requires a team effort with a close working relationship between sales, marketing, engineering, and manufacturing. A team including the president of the company, and representatives from engineering, manufacturing, finance, marketing/sales, and material management should be formed to develop the MPS.

The team should establish the acceptable delivery time for products, determine the impact of engineering changes and new product introductions on the forecast, and develop the forecast for products at the highest product level. The team should clearly understand the impact of lead times on the master schedule. For example, the MPS for a prod-uct with a 16 week lead time cannot have any changes within that time window without creating excess inventory or unavailable products. This is a critical point that is often misunderstood by most sales and marketing people.

The staff should also consider backlog when planning the MPS. Many companies strive to ship their backlog as soon as possible, depriving them of the ability to smooth out order fluctuations. This planning requires an understanding of the customer's delivery requirements. The larger you can make your backlog, the more flexibility you can have in planning inventory and production runs. Of course, customer delivery requirements and competitive pressures will place constraints on your ability to use backlog to smooth out order fluctuations.

Make It or Buy It?

The decision to make or buy subassemblies can have significant financial and strategic impact on your company. Large, established companies such as Sun Microsystems, Apple Computer, and Hewlett-Packard regard manufacturing capability to be a significant strategic advantage, but even these companies purchase subassemblies for their products from subcontractors. You should consider where your strategic advantage is and develop your manufacturing operation to maximize it.

Board subassembly manufacturing requires a significant investment in capital equipment and human resources. Many subcontractors for board assembly can provide turnkey operation in which they purchase the parts, load and test the boards, and provide complete quality control. This alternative can be especially beneficial if you're using surface mount technology. Even the largest companies have found that it is very expensive to keep up with rapid technology changes. Subcontractors may also be able to buy parts cheaper than you because of their high volume operation.

The volume of products that is required to interest a contract manufacturer depends on the strategic value of your company and the value of the revenue stream. The strategic value may be a desire to participate in a particular industry. A product which has a very high labor content but lower volume may be of more interest to the contractor than a moderate volume, low labor content product because it maximizes the value added labor for the vendor. Many vendors are willing to grow with a new customer, but because their margins are extremely thin they strive to minimize their risk.

The larger contractors can provide lots of services during the design phase, including design guidelines and complete printed circuit board layout. The contractor will require good documentation to manufacture your boards, such as an accurate bill of materials (BOM), fabrication drawings, assembly drawings, and schematics for board test. Additional documentation that is desirable includes purchased parts vendor list, drill drawings, programmable array logic (PAL) library, and a "gold standard" (i.e., one that works perfectly) card. The larger companies also provide a high level of manufacturing expertise early in the design phase of the product and a well developed quality assurance department after production commences. Premier companies such as Hewlett-Packard have found that the defect rate from the better contract manufacturers is lower than their own performance because of the contractor's total focus on manufacturing. Defect rates as low as 160 parts per 1,000,000 have been reported by Solectron, a Silicon Valley contractor and winner of the 1991 Malcolm Baldridge Award.

Product testing is another area that requires major attention in the start-up and production phase of your company. Fast board testing capability often requires a heavy investment in automated test systems and significant production engineering costs to develop test software and fixtures. The investment to develop this capability may be better spent on marketing, sales, or product development. There are automated test houses that provide services on a contract basis. Be sure to include the vendor in the design phase of the project so that the end result will be easily tested on their equipment.

Most products consist of more than printed circuit boards and assemblies. Sheet metal, plastic parts, castings, and machine parts are usually part of the finished product and often represent much of the design and manufacturing cost. Fabrication of these parts involves extensive use of heavy machinery such as numerically controlled punches and plastic molding equipment. Consequently, these parts are usually subcontracted even by the largest manufacturers. Vendors of these types of parts should also be involved in the design phase so that the parts are easily produced with low reject and scrap rates during production.

Some products can be completely produced on a subcontract basis. For example, one company is producing a computer mouse that is entirely constructed, packaged, and shipped to designated distributors by the subcontractor. Contract manufacturers are continually looking for new ways to increase their production volume and value-added content for their customers. Manufacturing capability has strategic importance; decide how "make or buy" will fit into your company's plan.

Getting It Together

How do you develop a manufacturing organization that can produce the highest quality products at the lowest possible price? Much has been written about Japanese manufacturing capability. The experience of Trimble Navigation illustrates the point. Trimble licensed some of its technology, which included a low cost satellite receiver design, to Pioneer to incorporate into car navigation systems in Japan. Trimble's engineers thought that they had really value engineered the design. They were in for a surprise; within six months, Pioneer showed them a new design that used no new technology, but reduced the manufacturing cost of the receiver by 50%.

The philosophy that seems to drive the system within Japan and also works for some American companies is the following definition of just-in-time (JIT) manufacturing:

"A philosophy of reducing waste and continual improvement"

But what exactly is JIT? A fledgling company can hardly know its sales volume accurately enough or have the muscle to get vendors to deliver components or assemblies to the loading dock twice a day and within 15 minutes of a designated time. However, JIT really addresses execution of plans and ways to continuously improve operations. Procedures would seem foreign to the concept of continuous improvement, but in fact they are simply a method of documenting the improvements that are made. The elimination of waste is not simply reducing scrap, but removal of any process or action that does not add value to the product or knowledge of the company's performance. These concepts are implemented through an effective manufacturing operation which includes high performance in documentation, materials, traffic, assembly, test, and quality.

Documentation

Many companies often focus on the processes that make up the manufacturing cycle, but documentation controls the operation. In order for work teams to function efficiently, and planning to be effective, documentation must be perfect. In Chapter 10, we extol the benefits of manufacturing resource planning (MRP), which can be used to develop material and manpower requirements for a given production schedule. But without pristine records and documentation, MRP will only enable you to purchase the wrong parts for the wrong assemblies. The bills of material, drawings, and process sheets must be accurate and up to date for manufacturing to meet its goals. High priority must be given to the control and maintenance of these vital records.

Materials

Direct material is a high cost item for high-tech manufacturing companies, often over 25% of sales. Consequently, it deserves a lot of attention from management. In addition to the original part cost, there are also overhead costs such as interest on inventory, administration of purchasing, and warehouse costs. Parts proliferation can significantly add to indirect costs by reducing the size of volume purchases, increasing the number of purchase orders, and increasing needed warehouse space. An approved parts list that is jointly developed by engineering and manufacturing can greatly reduce part proliferation.

Engineering influences the ability of the materials department to acquire inventory in a timely fashion. Product structures that are not well designed and processes that cannot be readily performed by one vendor will increase the lead times of parts. Long lead time parts should be avoided if there are other alternatives available, even if they cost more.

A useful technique for managing inventory and product availability is to develop product profiles of all products. This is illustrated in Figure 8-2.

Figure 8-2 shows that the lead-time to produce a model 540A can be reduced from 12 weeks to 5 weeks, with an inventory of $411.12 worth of components. If this level of commitment is too high, the lead time could be reduced from 12 weeks to 8 weeks with an inventory of $260.56 of components. Minimizing the lead time of the components of your product can significantly reduce your inventory requirements and increase your planning flexibility. Of course, the lead times in your database must be very accurate.

Effective management of inventory levels is important for any company, but it can be critical for high-tech companies. For example, the semiconductor industry is notorious for having extremely long lead times for popular products and obsoleting products that don't attain adequate production volumes. For example, at the end of 1989 Intel had a monopoly on the popular 386SX microprocessor that was being used by all manufacturers of high-end IBM compatible personal computers. Many equipment manufacturers could not get an adequate supply and were at Intel's mercy. A safety stock level should be maintained to provide a buffer for production, particularly for critical, long lead time items that are hard to procure.

Fortunately, all items aren't in the category of the 386 microprocessor. Many common components are available on very short notice. A useful technique for managing inventory levels is to segregate parts into three categories like this method:

A—Parts that account for 60% of annual dollar volume

B—Parts that account for 30% of annual dollar volume

C—Parts that account for 10% of annual dollar volume

"A" parts are typically purchased on blanket orders for scheduled delivery as required. "B" parts are typically bought quarterly or semi-annually, and "C" parts are purchased semiannually or annually. This

Product Profile 540A-1

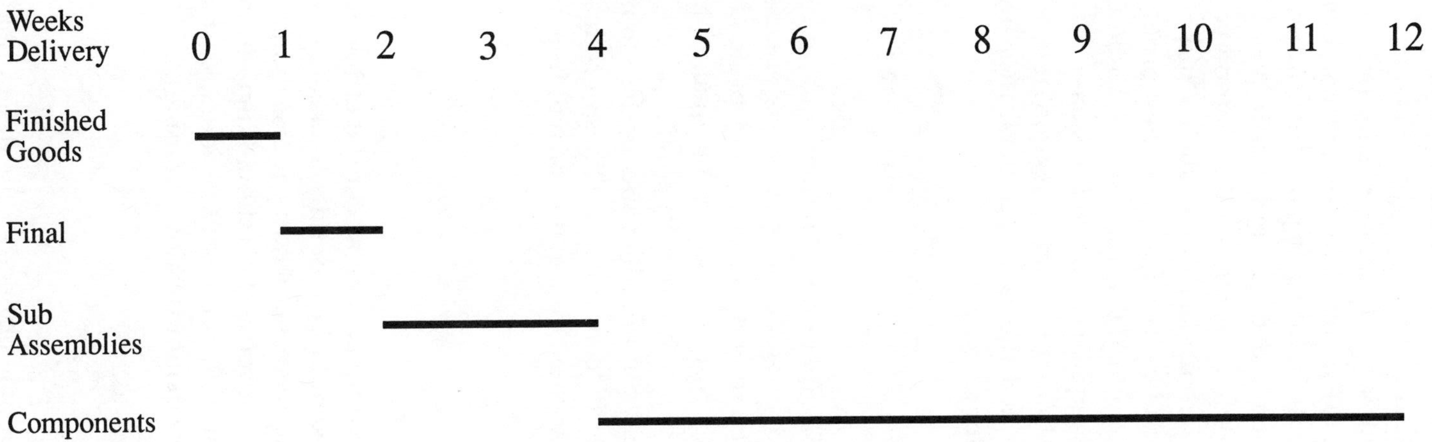
Weeks Delivery
0
1
2
3
4
5
6
7
8
9
10
11
12
Finished Goods
Final
Sub Assemblies
Components

Weeks Delivery	0	1	2	3	4	5	6	7	8	9	10	11	12
Materials	18.85			1.76	8.95	55.46	41.51	53.59	23.00	175.26	12.35		49.95
Labor	6.00		45.00		26.50								
Labor+OH	23.52		176.40		103.88								
Mat'l+ Lab+OH	42.37		176.40	1.76	112.83	55.46	41.51	53.59	23.00	175.26	12.35		49.95
Cum Total	744.48	702.11	702.11	525.71	523.95	411.12	355.66	314.15	260.56	237.56	62.30	49.95	49.95

Figure 8-2: A product profile lets you evaluate the time and materials necessary to manufacture a certain product.

technique reduces administrative time, assures that low dollar volume parts are available when needed, and focuses management time on parts where you can save the most money.

Poor part quality can also impact your ability to ship products. A quality program will assure that the material you receive conforms to your specifications, particularly the fabricated parts such as printed circuit boards, sheet metal, machined parts, and plastic parts. This may be done on a sample or 100% inspection basis. A good vendor certification program can save money here by reducing inspection costs. Material in the pipeline can be a particularly vexing problem if your company is trying to implement JIT at the level of Apple Computer. Apple discovered in a sample test of CRT displays that the phosphor was the wrong color. Upon further investigation, they found that the next 6 months' supply had the same problem; they had to shut down the production line for six weeks.

Assembly

JIT also means training and empowering employees to make improvements in manufacturing operations. How products are assembled, tested, and materials moved to and from the work place are all part of the process of reducing cost and improving quality. Many modern factories are organized into work cells where production teams have complete responsibility for a product and are free to make changes in the processes required to produce it. These groups are flexible and cross-trained so that they are capable of building other products as demand shifts.

Traffic

How material moves in the company can affect the efficiency of the organization. Material flow should be carefully tracked to assure that there is adequate information to know what transactions have taken place and where material is located. This should include stock-room inventory, shop floor material, returns to vendors, customer shipments and returns, and sales demonstration equipment.

Test

Product testing should be carefully planned in the design phase of the product. The goal should be to design tests that assure the quality of the product, but are not overkill. Quality cannot be tested into a product; it must be designed in and built into the manufacturing process.

Quality

A new level of quality consciousness was born in the 1980s when Japanese manufacturers introduced product after product with more features and unprecedented levels of consistent quality. As the 1990s began, even the vaunted German luxury car market was under assault by the "Lexus" model from Toyota, which offered outstanding performance and quality at prices that were two-thirds of the comparable German offerings. In Japan, continuous pursuit of perfection is the only acceptable goal. How can you compete with this level of commitment to quality?

First, you must involve all of your employees in the process to improve the quality of your goods and services and empower them to make improvements. Second, a measurement system should be developed to provide feedback on how the organization is performing. For example, knowing why a particular assembly or product fails a test can provide insight into the necessary remedies for the problem. Without the measurement system there is no way to know where to look for the solution. Finally, your people have to feel that they have the time to "do it right the first time." Unrealistic deadlines are often placed on projects and production schedules, producing an attitude that it is okay to do a task over rather than to take the time to do it right. Workers at the lowest level often recognize problems that cause poor quality and raise costs, but they cannot get management to make changes.

Can We Do It?

Modern computers and software have made this question less formidable than it once was. Manufacturing resource planning software was introduced in the early 1970s and is in wide use in industry today. These MRP software packages can provide a detailed description of the materials and labor that you will need to satisfy a master production schedule, down to how many 100 Ω resistors will be needed in the thirteenth week of the production schedule. If the MPS is wrong, you will have the wrong parts—an obvious observation, but one that is overlooked every day.

MRP won't tell you if you have adequate plant, equipment, and labor available to meet the MPS; it only concerns itself with the number of labor hours required to complete a process and the material necessary for the process. Very few companies implement the full capability of the medium to large scale MRP systems that are available today. Companies can operate the system software and hardware, but the records and procedures required to establish and maintain the integrity of the database for the program are not in place. A company must have pristine records for MRP to be useful. This implies that all the transactions that people make during the workday are accurately recorded. This is the Achilles heel of modern information management systems. Your company should strive to be in the top 10% of class A MRP systems by meeting the following requirements:

▼ *Use the system to manage the business.*
   The reports of product availability and cost are accurate enough to reliably quote delivery and make pricing decisions.

▼ *BOM accuracy of at least 99%.*
   When kits are pulled for assemblies, you get the right parts.

▼ *Stock status accuracy of at least 95%.*
   Unexpected shortages do not show up when kits are pulled.

▼ *No expediters.*
   Vendors deliveries are reliable enough that there is no need for expediting to get late deliveries corrected. Stock is available on time so you can shorten lead times.

▼ *Materials planning.*
   MRP develops demand by part number based on the MPS. The program will balance part availability against stock on hand and purchase orders that are due to be filled. The better programs provide exception reporting so that materials personnel can concentrate their efforts on the problem parts. This tool can greatly increase the efficiency of the materials group. Handar implemented the material side of MRP as a $1,000,000 company. The company estimated that twice as many staff would have been required to handle the material transactions without the system.

▼ *Capacity planning.*
   MRP can also address the labor side of the production operation by planning the amount of labor required to implement the MPS. The labor is planned by work center so that a detailed

breakdown covering assembly, test, inspection, and any other processes required to produce the product are available.

How Are We Doing?

Timely reporting on the performance of the manufacturing operation is critical to the successful management of your company. Many of the quarterly earnings surprises reported by publicly traded manufacturing companies come from unexpected write-offs of obsolete inventory or higher than anticipated production costs. Track the rate at which shipments are being made during each month and make every effort to avoid the month end shipment crunch—the familiar "hockey stick" effect.

Develop meaningful reports on the quality of the products being produced. For example, monitor the failure rate of products in process to provide insight into the performance of the internal quality assurance program. Warranty, dead on arrival (DOA), and return rates of products shipped to your customers should also be monitored. For example, products can be randomly selected from the shipping dock, set up by a shipping clerk, and turned on to verify performance. An on-time delivery report that shows the percentage of the time that products were delivered to customers by the promise dates is another good measure of the efficiency of the operation and your commitment to customer satisfaction.

Reports on the usability and value of inventory are extremely important since inventory is such a high exposure item for high-tech manufacturers. Monitor inventory trends and purchasing commitments to be sure that material in stock is not growing at a faster rate than shipment projections. If you are in a high inflationary environment, the parts costs should be carefully tracked so that unfavorable variances can be adjusted in a timely fashion.

Keep track of the gross margin (the difference between sales price and manufactured cost) of the products you produce by product. Track labor and material costs and make adjustments in pricing or manufacturing where margins are under attack because of cost escalation. Variances between labor and material standards and actual cost should be measured and the cost of goods sold adjusted monthly to reflect the differences. If the adjustments are not made regularly, management may have a big surprise at the end of the quarter or year when new labor and material standards are set.

Electronics Manufacturing in the 1990s

This chapter has presented most of the manufacturing concepts that you will need to lead your company through the 1990s. Some of the observations on future trends by experts in the field are as follows:

Computer Integrated Manufacturing (CIM)

Islands of automation will begin to be connected in the 1990s. It will be a slow process because all of the elements for CIM technology are not yet in place. At the time this book was being written, IBM was the only company to have implemented nearly all of the technology developed through 1989, and at a cost of $22,000,000,000. Traditional accounting will change to permit the implementation of CIM so that the cost to the total enterprise is considered in the payback analysis.

Product development

Product cycle times are anticipated to drop from 24 months to 12 months as competition heats up. Close coordination between design, materials management, and manufacturing will play a key role in meeting this challenge.

Processes and equipment

At the time this book was written, surface mount technology (SMT) was projected to be used on over 70% of all printed circuit boards by the early 1990s. SMT will require completely automated insertion lines and soldering equipment, and will be done by flexible equipment that can easily be reconfigured to handle different products.

Decision support systems

Artificial intelligence will be employed to create systems that "learn" from their past mistakes. Expert systems will be extensively used to troubleshoot products both in the factory and at the customer site. This technology, along with more powerful, affordable computer systems, will provide management with powerful decision making tools. For example, a complete regeneration of an MRP forecast may be run in minutes, where older systems may have taken up to 12 hours. This capability will provide unprecedented opportunity for managers to run "what if" simulations.

JIT

JIT techniques will be pursued with increased intensity. A survey of four Japanese manufacturers conducted in 1989 by Schwafel & Associates, manufacturing consultants, showed the results summarized in Table 8-1:

TABLE 8-1

Company	Program Duration	Inventory Reduction	Throughput Time Reduction	Labor Productivity Increase
A	3 years	45%	40%	50%
B	3 years	16%	20%	80%
C	4 years	30%	25%	60%
D	2 years	20%	50%	50%

Clearly, JIT has tremendous promise for improving the profitability of manufacturing enterprises and will be of increasing importance in the 1990s.

CHAPTER 9

Quality is Not a Handbook, It's a Philosophy

*T*he term "quality control" often creates an image of a staff of inspectors armed with quality control standards and manuals overseeing every phase of a manufacturing operation. The Japanese, with the help of W. Edwards Deming, have demonstrated that inspectors and handbooks are not the way to achieve high quality in products and organizations. Quality products and organizations are built by setting high standards, measuring performance relative to the standards, and involving the people doing the work in the process. (In Appendix I, we've summarized Deming's 14 basic principles.)

Quality standards for a product or for an organization are not easily established. If standards are set too high, employees can become discouraged, believing the standard is unreachable, something like a search for the Holy Grail. If the standard is too low, then product or service quality can be unacceptable. The following definition of quality overcomes these problems:

> *Quality is the search for zero defects in all areas of the organization. The goal is zero defects and progress is measured by the rate of decrease in defects in all areas of the organization.*

This definition provides a way to celebrate small successes along the trail to perfection.

A successful quality assurance program requires a commitment from you and your management team in both words and actions. Each department should develop its own quality standards with a measurement system to show how they are performing. Get involved with the people in the trenches doing the work whether in manufacturing, sales, or accounting. In addition to internal feedback, solicit comments from your customers. A call to a customer to find out how a product works, how the service has been, or if there is anything lacking in your support will provide valuable insight as to how the company is performing. This is total quality control (TQC) for the whole company.

A quality assurance manager that reports to the president should be designated for the company as soon as possible. This person will usually have more than one responsibility in the early stages of the company. The quality assurance responsibility should be to develop standards, train people, and measure performance. Attention to these details may seem inappropriate in the start-up phase of the company because everyone has so much to do, but if quality is not addressed, a "no quality" culture may develop that will haunt the company for a long time. Poor quality of a first product can be disaster for a start-up.

Quality in Manufacturing

Product defects usually come to mind when quality is discussed. This is no surprise since the results of poor quality are so obvious when products are returned and there's a lot of scrap in production. How can you avoid these problems?

A system of self-inspection using defined standards should be developed with the help of the people doing the work. There are books available from companies such as 3M which establish manufacturing standards for most processes. All manufacturing departments should be included in the program.

The quality of the parts supplied by your vendors also plays an important role in product quality and production cost. A vendor certification program should be established to insure high quality and reasonable cost of the purchased parts. The program should measure the performance of your vendors in pricing, on time delivery, defect rates, paperwork performance, and facility quality. If you can't develop your own guidelines, borrow them from another company or hire a consultant to help develop them. More than one vendor should be qualified to assure adequate price competition, but don't purchase simply on low cost—you may pay with lower quality or higher rework costs. Your vendors should be partners in your future success.

Testing is another important ingredient in the quality of manufactured products. Develop a comprehensive test plan for each product that is consistent with its use and specifications. For example, Campbell Scientific of Logan, Utah manufactures environmental data collection equipment that has to work reliably in weather extremes throughout the world. The company adopted a policy of 100% operational temperature testing from −20° to +50° centigrade. The company's products have a reputation for the high reliability.

However, it's also common for many companies to excessively test their products. In 1988, *Electronic Business* magazine and Coopers & Lybrand conducted a survey of the quality assurance policies of 144 companies and found that 75% could make significant cost reductions because of redundant testing policies. Even though these companies were quite familiar with modern TQC techniques, they did not abandon their 100% inspection programs after implementing statistical quality control or vendor certification programs. The lesson here is to establish your testing plans carefully to control both quality and cost.

Most companies find that the "Betty Crocker" method of quality assurance helps reduce infant mortality in products—bake at 50° centigrade for 24 hours. In 1990, Apple Computer spent one hour assembling a Macintosh and 24 hours in a "burn in" operation to shake out infant mortality. Cycling the electrical power to the product on and off for short periods also adds stress and can ferret out weak components. This process increases product inventory, but the lower failure rates you will obtain will more than offset the the additional cost in inventory and testing.

How do you know how you're doing? Surveys that measure customer satisfaction with products are well known and extensively used. However, response rates are usually low. Handar came up with a method that provided a very high 67% return rate. The company developed a form letter that was personalized for each customer and signed by the president. A short prepaid response card was included for their response. Develop a method that gives you feedback on your performance.

Quality in Engineering

Quality in engineering begins at the design concept for a product. A well conceived design with consideration for maintenance, manufacturing, and ease of use will usually result in a very successful product. A poor design cannot be saved by lavish customer service attention or intensive testing in manufacturing. Products from these designs are called "dogs" by your manufacturing people and your customers.

The next step is implementation of the design. Use conservative design guidelines for part ratings and have adequate margins for component parameters that are critical to a product's performance. Don't use components beyond their stated operational range and don't count on unspecified performance that is critical to the successful operation of a product. The probability is 100% that the part will eventually fail to meet your needs when the parameter drifts out of its normal range.

High temperature is another enemy of reliable performance in most electronic hardware. Observe component ratings and strive to minimize high temperatures throughout the interior of the product. Component derating guidelines can be very useful in improving product reliability. The U.S. Department of Defense has guidelines that can be useful for developing a commercial standard. Quality assurance consultants are also available to help develop guidelines.

A standard parts list should be developed for new designs. Most engineers prefer to use the latest component in new designs, but this should only be done if there is a demonstrated improvement in performance or cost that is important to the customer. Parts proliferation needlessly contributes to a higher cost structure and lower inherent quality by increasing both the number of vendors to be dealt with and the number of different parts to be purchased. For example, we know of a $25,000,000 company that allowed each engineer to select the type of screws used to assemble their product. This company had more than four different types of #4 fastener screw. The company often dealt with as many as ten different vendors for the same part.

A product environmental testing policy should be established for production qualification. The parameters will depend on the product type; office equipment needs less stringent tests than outdoor equipment. The policy should include temperature tests, vibration and shock tests for shipping, electromagnetic compatibility, safety, and humidity tests. The U.S. Department of Defense has guidelines for procurement of commercial equipment that can be used to develop a sensible program.

How do you measure the performance of engineering on quality issues? The number of engineering change notices (ECNs) that are written after a product is released to production is a good quality indicator for most designs. A balance should be maintained—if no ECNs are issued, the product probably spent too much time in engineering; if every assembly has an ECN, the job was probably poorly done. Remember, changes are more expensive the closer the product is to the customer. A major recall after one year's production can break a start-up company.

The role that research and development plays in the development of a new product can be compared to the training of athletes preparing for the Olympics. How fast they run the 100 meters in training doesn't count unless they can achieve the same performance in the real event. The primary responsibility of the engineering department is to develop innovative products that can be manufactured at a competitive cost. The research and development work is preparation for the main event, namely building the product. Research and development's mission in a manufacturing company is to produce quality documentation for the manufacturing department. The days when research and development could just throw a design over the wall to the manufacturing department and walk away from it are long past. In successful companies, research and development and manufacturing work as a team. The quality of the documentation package is one of research and development's contributions to the team effort.

Quality in Marketing

Marketing quality starts with a good understanding of your customer's needs, competitive situations, and market trends. This knowledge must be effectively translated into good product descriptions and specifications. If marketing did no other tasks well, your company would probably be successful. Customer acceptance of new products is a good measure of the quality of the marketing job.

High quality literature must also be developed to present the right image of your product and company. "High quality" does not necessarily mean "high cost." Creative graphic designers can often develop very good literature on a low budget, although this is not their preferred method. Will your literature make the right statement about your product and company?

A product may be well conceived, properly designed and manufactured, and have good collateral literature, but suffer from a poorly conceived promotional plan. Customers have to know about your products and that usually means a promotional campaign. The campaign should be effective and present a positive corporate image. A quality standard for marketing communications materials should be established to maintain a consistent image for all product promotions.

Quality in Sales

In basketball, a point guard is the player who has most of the ball-handling responsibilities. He or she is the player who calls plays on offense, calls defensive signals, passes the ball to teammates in scoring position—the point guard is like a coach on the floor. The salesperson is the "point guard" for your company. As one veteran salesperson said, "Nothing happens until somebody sells something!" This is where opportunity begins or problems start. An important part of your TQC program is to get the sales order right. Errors in order processing or selling the wrong product to a customer can wreak havoc with production schedules and customer satisfaction.

The timeliness, accuracy, presentation, and completeness of customer quotations make a statement about your company. Detailed proposals can be developed into standard packages using a desktop publishing system and can be tailored to individual customers to personalize the quote.

A telephone call is often the first interaction a prospective customer has with your company. One company president heard the telephone ring six times, at which time he answered it. He almost fired the next person he met because he remembered the early days of his company when the phone seldom rang. Train your employees to answer the phones.

Measuring Performance

You have to establish standards and measurement techniques to achieve your quality objectives. Statistical sampling methods and quality assurance procedures should be established in the manufacturing department using the latest methods that can be applied to your company. A complete description of statistical process control and quality assurance is beyond the scope of this book. You may want to hire a consultant to set up measurement systems and training for your company.

A brief management philosophy should be written and a concise handbook and appropriate training should be developed for your employees, particularly for the manufacturing operation. We don't advocate a handbook/inspector approach to quality, but your employees need to know the company's expectations and how to meet them.

What about other departments? Can performance standards be set for accounting, product support, and product support? Quality control is customarily considered to be the province of product manufacturing, but consider Domino's Pizza's claim that they will pay if the pizza delivery time exceeds thirty minutes. All departments should have some written performance guidelines that employees help develop and performance should be reviewed on a regular basis. This is what total quality control means.

The cost of implementing all of this quality stuff is often a concern for the start-up company. But if you want to have a world class organization, regardless of your size, the whole organization must be committed to TQC. You will save money and your customers will applaud.

CHAPTER 10

Maintaining Financial Control

N ew high-tech industries grow explosively compared to the evolution of most older manufacturing industries. Even in high-technology itself, the pace accelerated through the 1980s. For example, Sun Microsystems became a $1,000,000,000 company in less than a decade. By contrast, it took Hewlett-Packard nearly 30 years to reach the same level of revenues. Such rapid growth can often lead to lack of financial control of the company. You must have systems and procedures in place to manage and control the finances of your company when you open your doors for business.

In the start-up phase, tracking development expenses and monitoring progress against the budgeted development cost is all that is required. The requirements can often be satisfied by one of the inexpensive personal computer software packages that can manage check writing, accounts payable, and vendor tracking. A more sophisticated system will have to be in place when manufacturing begins.

You should select an outside accounting firm to help set up an accounting system, prepare tax returns, and provide outside reviews or audits of the books. Choose a national firm, such as Arthur Anderson or Price Waterhouse, if your plans are to grow at a very high rate and be a publicly traded company three to five years after start-up. If your growth projections are more modest, a regional accounting firm that is familiar with high-tech manufacturing companies can probably serve your needs at a lower cost. Some national firms have established departments to deal with smaller accounts, so shop around for this service. Whatever your selection, make sure that the firm can provide the services you will need, such as tax advice, financial consulting, profit sharing plans, international tax advice, and basic accounting consulting.

If you plan to deal primarily with the federal government, select an accounting firm that is familiar with government requirements and set up the accounting system to meet government auditing standards.

It is possible and permissible to keep separate books for government audits, but extra expense will be incurred.

One of the first items to be dealt with is payroll. There are many companies that provide payroll services such as writing payroll checks and making withholding tax deposits. Your bank can provide recommendations. Be sure that labor costs can be easily allocated to different job accounts on your internal accounting system. The extra effort required to post labor costs on your accounting system may cancel any savings that might be achieved with the service. There are personal computer software packages that can track payroll and calculate the withholding tax deposits. The process for making withholding tax deposits must be understood by your staff and done on time, because the penalties for a late or "under" deposit is 10% of the amount owed. You will receive a manual from the federal and state government shortly after you incorporate that describes how to make withholding deposits. If you don't receive these, contact the Internal Revenue Service and the appropriate agency in your state.

Accurate reporting is essential for the accounting system to be of any use to you. Educate your staff on the importance of accurate reporting of financial transactions for the company. For example, items that are incorrectly charged to a job number will distort the job costs. Profit on jobs will be misstated, leading to erroneous pricing in the future; this obvious observation is often overlooked by many companies.

After you start shipping your product, you will be faced with collecting money from your customers. This is one of the areas in business that is not fun. Some customers have a myriad of ways to forestall payment either intentionally or inadvertently. The first step is to make sure that your customers are worthy of receiving credit. You should establish a credit approval procedure so that you don't deliver products to customers that won't pay. Dun & Bradstreet reports are helpful for verifying credit as well as vendor and bank references. A standard package should be developed that is sent to all first time customers before a sale is made. Your bank or accountant can help you develop the package.

Some customers will always be difficult to collect from because of their payment policies. You may be thrilled by your first order from a large company only to find that payment stretches out to 120 days. Find out what the company's payment policy is before accepting a large order. It could turn out to be a financial albatross if payment is delayed. Many large companies augment their credit lines at the expense of small companies by stretching out payment.

Make sure that all of your shipping and invoice documents are in order, especially when dealing with large companies or government agencies. Any mistake in the billing or shipping paperwork is guaranteed to delay payment.

A good method for shortening receivable collection periods is to have a clerk call the customer ten to fifteen days after a shipment has been made to make sure the product was received in good working order and that all of the paperwork was in order. This is only practical with higher cost products, but can be very effective in facilitating receivable collections. As a last resort, you can always turn the collections over to a collection agency.

Are We Making Any Money?

The profit-and-loss statement (P&L) tracks profitability. Keep this in mind as you assign account numbers for the company. Break down cost areas so that adequate control can be maintained, but not so fine that the detail is overwhelming. Budget categories should not be less than 1 to 2% of a total budget. For example, if the marketing department has a budget of $300,000, then no category would have a budget less than $3000 to $6000. Table 10-1 illustrates a sum-mary P&L statement for Trimble Navigation in 1985, and Table 10-2 is an example of a hypothetical department expense statement.

TABLE 10-1

Trimble Navigation Expense Statement

(all amounts in thousands except per-share amounts)

	Dollar Amount	% of Total
NET REVENUE:		
Product sales	$4402	87.1
Technology licenses	650	12.9
Net Revenue	5052	100
OPERATING EXPENSES:		
Cost of product sales	1526	30.2
Research and development	1205	23.9
Sales and marketing	882	17.5
General and administrative	687	13.6
Total Operating Expense	4300	85.1
OPERATING INCOME	$752	14.9
Interest expense	194	3.8
Income before taxes	558	11.1
Provision for income taxes	281	5.6
NET INCOME	$277	5.5
NET INCOME PER SHARE		$0.03

AVERAGE COMMON SHARES
OUTSTANDING: 7,951,000

TABLE 10-2

Marketing and Sales Expenses
(all amounts in thousands)

MARKETING AND SALES EXPENSE	Amount	% of Sales
Product sales	$440.2	
Salaries	400.2	9.1
Commissions	300.0	6.8
Travel	45.5	1.0
Payroll taxes	30.9	0.7
Office supplies	5.3	0.1
Literature and printing	21.5	0.5
Trade shows	15.7	0.4
Depreciation	6.5	0.2
Vacation, holiday, and sick pay	37.9	0.9
Employee benefits	18.2	0.4
TOTALS	$881.7	20.0

The P&L found in most annual reports has no account numbers associated with the categories. However, you will want account numbers assigned to these categories shown in Table 10-2. There is no magic formula for choosing account numbers; use a system that makes sense to you, your accountant, and your computer.

Budgets for departments, projects, and special jobs are a must for maintaining control of the cash flow in your company. A wise person once said, "If you don't know where you are going, how will you know when you get there?" So it is with budgets and P&L statements; without budgets, you will not know if you are underspending or overspending in various categories. Initial budgets can only be based on past experience. Do the best you can especially on the development project cost. You can get help from outside vendors for estimates on prototypes for fabricated parts and other services. If you do not have experience in this area, hire a consultant or get help from your accountant.

The budgeting process will become more detailed as your company grows. An investment policy should be developed for all areas of the company as part of its growth strategy. Should you spend 23% on research and development as Trimble Navigation did in 1985, or is 10% the correct number for your industry and market? How much should marketing and sales be? For example, Apple Computer spent 23% of revenues on this item in 1989, but only 8% for research and development. Only you can determine these numbers; make them part of your strategic planning process.

The more mature company will start the budgeting process with an estimate of the new year's order and shipment revenue. The num-bers are usually negotiated between the sales organization, manufacturing, and top management. After the order and shipment numbers are established, individual department expense levels can be established on a "top down" basis by using your investment criteria for each area. For example, if total shipments were to be $5,000,000 and research and development investment is 10%, then the top down budget for research and development would be $500,000. Zero-based budgeting where each department is required to justify next year's budget from a "bottom up" calculation is a good disciplinary exercise. This technique can reduce inefficient spending "creep" where expenses increase to devour the available dollars.

The break-even point is an important financial measure of company operations to be considered in the budgeting process. This is calculated by reducing shipment revenues to the point at which the operating profit becomes zero. Build a profit-and-loss model which has both fixed costs and costs that vary with shipments. Examples of fixed costs would be rent on the building, lease payments on equipment, or indirect labor salaries. Variable costs would be direct material, direct labor, and shipping costs. This technique will compute the shipment level to break even on a monthly basis and can easily be implemented on an electronic spreadsheet.

The P&L statement and its associated budget are the tools for monitoring your company's profitability. They provide a picture of its performance over a specific time interval, such as the most recent year of operations. They do not predict bankruptcies. This requires careful analysis of the balance sheet and the cash flow statement, which is discussed later.

How Healthy Is Your Company?

The balance sheet is a snapshot in time of the financial health of the company. Many companies have had profitable operations, but have gotten into trouble because of a weakened balance sheet and a lack of cash to pay the bills. Table 10-3 is a summary balance sheet for Trimble Navigation on December 31, 1989, the last complete fiscal year report before its initial public offering in July, 1990. The corresponding P&L statement is shown in Table 10-4.

TABLE 10-3

Trimble Navigation
Consolidated Balance Sheet
(all amounts in thousands)

ASSETS		LIABILITIES AND SHAREHOLDERS' EQUITY	
Current Assets:		Current Liabilities:	
Cash and equivalents	$471	Note payable to bank	$3,400
Short-term investments	44	Accounts payable	3,255
Accounts receivable	8,225	Accrued compensation	911
Inventories	5,182	Other accrued liabilities	394
Other	452	Income taxes payable	167
Total Current Assets	14,374	Capitalized lease obligation	454
		Total Current Liabilities	8,581
Property and Equipment, At Cost:			
Machinery and equipment	5,388	Convertible Subordinated Notes:	3,324
Furniture and fixtures	605		
Leasehold improvements	274	Capitalized Lease Obligations:	1,095
Subtotal	6,267		
Less accumulated		Shareholder's Equity:	
depreciation	(2,531)	Common stock	6,667
Net Property and Equipment	3,736	Notes receivable	
		from shareholders	(1,179)
Other Assets:	497	Retained earnings (deficit)	119
		Total Shareholder's Equity	5,607
TOTAL ASSETS	$18,607		
		TOTAL LIABILITIES AND SHAREHOLDER'S EQUITY	$18,607

The retained earnings of $119,000 are the cumulative profits from date of incorporation. They comprise a loss of $193,000 in 1988, the profit of $500,000 in 1989, and a stock repurchase of $188,000 in 1989.

An analysis of the P&L statement and the balance sheet of Trimble as a complementary pair of financial statements will illustrate how a company's financial health is measured. Both are important to assess the financial condition and efficiency of a company. The analysis can be carried out with a series of ratios which can be classified as follows:

TABLE 10-4

Trimble Navigation
Summary Profit and Loss Statement
(all amounts in thousands)

Net Revenues:	
Product sales	$27,860
Technology licenses and research	3,993
NET REVENUE	31,853
Operating Expenses:	
Cost of product sales	12,326
Research and development	8,616
Sales and marketing	5,885
General and administrative	3,370
TOTAL OPERATING EXPENSE	30,197
Operating Income:	1,656
Interest Expense:	737
Income Before Taxes:	919
Provision for Income Taxes:	419
NET INCOME	$500

▼ *Liquidity ratios*, which measure the firm's ability to meet its maturing short-term obligations.

▼ *Leverage ratios*, which measure the extent to which the firm has been financed by debt.

▼ *Activity ratios*, which measure how effectively the firm is using its resources.

▼ *Profitability ratios*, which measure management's ability to control costs as shown by the returns generated on sales and investment.

Liquidity Ratios

These ratios are quick tests which measure the company's ability to meet its short-term financial obligations. A detailed cash flow analysis is necessary to provide a more complete picture of the situation and is discussed later in this chapter.

▼ *Current Ratio*

The current ratio is computed by dividing the current assets by the current liabilities. Current assets are any assets which are easily converted into cash in a short time, such as receivables, short-term notes payable (less than one year), and inventories. Current liabilities are the current part (amounts due less than one year) of long-term debt, short-term revolving debt, taxes, accounts payable, and other accrued expenses due within one year. Wages often show up as current liabilities because of payroll cutoff dates not corresponding to the end of the financial reporting period or accrued vacation liabilities that could be realized in the current year. The current ratio for Trimble is calculated as follows:

Current ratio = current assets/current liabilities

$14,374/$8,581 = 1.7

This ratio indicates a good degree of liquidity for Trimble. In 1990, typical ratios for large electronics companies such as Digital Equipment Corporation and Hewlett-Packard ranged from 2.7 for DEC, which was highly liquid, to 1.5 for H-P. Most publicly held high-tech companies have ratios that range from 2.2 to 2.5. High liquidity for a small privately held company makes it easier to raise debt financing from commercial banks.

▼ *Quick Ratio*

The quick ratio is calculated by deducting inventories from the current assets and dividing the remainder by current liabilities. Inventories are typically the least liquid assets and most likely to be written down in the event of a liquidation. The quick ratio for Trimble is calculated as follows:

Quick ratio = (current assets-inventories)/current liabilities

$9191/$3294 = 2.8

This ratio also indicates a very high liquidity for Trimble since in a liquidation the company has the creditors covered by over 180%. The typical publicly held electronics company has a quick ratio in the range of 1.0 to 1.5. This ratio can be as large as 2.0 for highly liquid companies such as DEC and Apple Computer were in 1990. Companies that have recently gone public are usually highly liquid because of the large amount of cash received in the public offering.

Leverage Ratios

These ratios measure the funds supplied by the company's owners (stockholders) as compared to the company's creditors (suppliers, banks). The creditors look to equity to protect their interests in a liquidation; the owners want to maximize their return through leverage (debt and credit) because this limits dilution of the owner's interest in the company and increases the return on shareholder equity (ROE). However, the debt leverage can work both ways. If the return on assets (ROA) is 20% and debt costs 10%, then there is a 10% differential that accrues to the company's owners. If, on the other hand, interest rates rise to 15% and the ROA falls to 10%, the differential of 5% must come from the owner's share of total profits. In the first case leverage is favorable to the owners, in the second case it is unfavorable.

Leverage can provide unpleasant surprises in a recession or if unforeseen business problems arise. A highly leveraged company that encounters a shortfall in shipments may not be able to meet its interest payments and may face action by its creditors. The suppliers can stop delivery of needed material to meet shipments and the bank could call outstanding notes, forcing the company into bankruptcy. The upside of leverage is that when times are good, the company gets to ride on the creditor's money and can earn a higher return on equity.

Leverage can be analyzed in two ways: on the balance sheet and from the P&L. The first provides insight into the likely outcome in a liquidation proceeding, while the second evaluates the probability of that event by looking at the coverage of fixed financing charges by earnings.

▼ *Debt to Total Assets*

This ratio (the debt ratio) measures the percentage of total funds provided by the creditors. Owners like to keep this ratio high to maintain control of the company—if debt financing can't be raised, more stock must be sold. Creditors like to keep the ratio low to minimize their exposure and keep the pressure on the owners to perform. If the ratio becomes too high, the owners have little exposure to a business failure. The debt ratio for Trimble can be calculated as follows:

Debt ratio = total debt/total assets

$9676/$18,607 = 52%

The total debt for Trimble was calculated by adding the capitalized lease obligations of $1,095,000 to the current liabilities. The convertible subordinated notes of $3,324,000 were not included because they can be converted to common stock and come behind other creditors in a liquidation proceeding. If this debt is included the ratio is 70%. The company is highly leveraged, which is not uncommon for a high-tech company that is planning to raise money on a public stock offering. Debt is cheaper financing for a closely-held company than selling equity because the stock is not liquid. Industry averages are usually used to gauge the exposure of the creditors, but they are not too useful for young high-tech companies.

▼ *Times Interest Earned*

This ratio, sometimes called "interest coverage," measures the ability of the company to cover interest payments with current earnings. It is calculated by dividing earnings before taxes and interest by the annual interest charges. If a company cannot service its debt, it may be forced into bankruptcy by its creditors. Pretax income is used in the numerator because taxes are calculated after the deduction is taken for interest. The times interest earned for Trimble is calculated as follows:

Times interest earned = operating income/interest expense

$1656/$737 = 2.2

The coverage of 2.2 times is rather low considering that Trimble's operating profit margin is only 5.2%, and that all of the profit probably came from technology licensing of $3,990,000. This ratio and the fixed charge ratio should be evaluated by considering the stability of the company's profits and cash reserves so that a business downturn does not put the company into a default position with its creditors.

▼ *Fixed Charge Coverage*

This ratio is similar to the times interest earned ratio, but the annual lease obligations of the firm are also included. The annual lease charges are added to the operating income in the numerator and to the interest charges in the denominator. The fixed charge coverage for Trimble is calculated as follows:

Fixed charge coverage = (operating income + lease)/(interest expense + lease)

$2110/$1191 = 1.8

The low coverage number indicates that the company is highly leveraged considering the volatility of its profits.

Activity Ratios

Activity ratios measure how effectively the company uses its financial resources to generate sales. The assumption is that a company that is not effectively using its assets is a candidate for problems such as a cash flow shortfall. For example, accounts receivable that are too large compared to the company's sales rate may become uncollectible or force the company to borrow money at a high interest rate to meet current cash flow requirements.

▼ *Inventory Turnover*
  This ratio is sometimes calculated using annual sales in the numerator and average annual inventory in the denominator (the average of beginning year inventory and year end inventory). A more common practice, which we prefer, is to use the cost of goods sold in the numerator. This ratio indicates how effectively a company is managing its inventory and the level of exposure a company may have to obsolete inventory, a critical consideration in fast moving high-technology businesses. The inventory turnover for Trimble is calculated as follows:

  Inventory turnover = (1989 cost of goods sold)/((1988 inventory + 1989 inventory)/2)

  $12,326/$4496 = 2.7   = "# turns"

  This is a relatively low ratio as compared to Apple Computer, which turned its inventory 5.8 times in 1989. The ability of the company to increase this number depends on the efficiency of its purchasing and manufacturing department and its product mix. A company that is in a high volume standard product busi-ness should turn its inventory six to ten times per year, whereas a small company with several low volume products may only generate three or four turns per year.

▼ *Average Collection Period*
  This ratio is also called day's sales outstanding (DSO). It provides insight into the quality of a company's accounts receivable and its ability to collect on goods that have been sold. The company's sales terms also affect this number since a short credit period should reduce the ratio. A rising trend in

this ratio may indicate that the company is not collecting its bills on time and should pay more attention to its collections procedures. It may also indicate quality problems because customers are withholding payment.

The ratio is calculated by dividing the company's annual sales by 365 to compute sales per day. This number is divided into the company's receivables to compute the collection period. The average collection period for Trimble is calculated as follows:

Sales per day = annual sales/365

$27,860,000/365 = $76,300/day

Average collection = accounts receivable/sales per day

$8,225,000/$76,300 = 108 days

Trimble's average collection period is quite high. The company's sales terms are net 30 days so a more reasonable number would be 45 to 60 days. The high number may indicate a need to review the collection process, the credit terms, and the business arrangements with its dealers.

A related financial tool is the accounts receivable aging schedule which shows the amounts outstanding by age. The amounts are typically grouped in brackets of days from 0-30, 31-60, 61-90, etc. A healthy situation for a company whose payment terms are net 30 is to have greater than 80% of its receivables age less than 45 days.

▼   *Total Asset Turnover*

This ratio measures how efficiently assets are used to generate sales. Total assets include property and equipment net of accrued depreciation, cash and equivalents, accounts receivable, and inventories. The ratio is computed by dividing the company's annual sales revenue by its total asset base. For Trimble this ratio is calculated as follows:

Total asset turnover = total revenue/total assets

$31,853/$18,607 = 1.7

By comparison, Digital Equipment Corporation and Apple Computer have turnover ratios in the range of 1.2 to 1.9. This ratio and the profitability show how fast the company can grow without additional debt or equity financing.

Profitability Ratios

These ratios measure how effectively management controls expenses and costs.

▼ *Profit Margin on Sales*
The profit margin on sales is computed by dividing the net profit after taxes and interest by total sales. The profit margin for Trimble is calculated as follows:

Profit margin = net profit after taxes/net revenues

$500/$31,853 = 1.6%

This is a very small profit margin compared to established high-tech companies, which normally earn 5% to 7%. The low profit can be attributed to the high growth rate that the company is financing. By comparison, Novell, a high growth computer networking company, had after tax margins of 12% at the time of its public offering in 1985.

It's useful to analyze some possible scenarios that would significantly alter the profit picture for Trimble. First, the company had revenues of $3,993,000 for business licenses in 1989, with probably no additional incurred costs. If these licenses had not been sold, the company would have incurred an operating income loss of $2,300,000.

The average collection period for receivables is an area that might be improved. If Trimble could reduce its average collection period from 107 to 60 days, the company would raise about $3,600,000 in cash which would retire the short-term bank note of $3,400,000 and put an additional $200,000 in the bank. The company borrowed the money at 12% interest, which incurred $408,000 of interest. Retiring this note would have reduced interest expense to $329,000, and raised the net income to $726,000, or 2.3%. The company would also have gained an extra $426,000 in cash.

The company's expenses as a percent of sales revenues could also be reviewed. For example, it spent $8.6 million on research and development which is 27% of net revenue and 31% of product sale revenue, both extraordinarily high amounts even for a high-tech electronics company. If the company could reduce its expenditures on R&D to 15%, the net income would soar to 8% to 10%. The ability to do this depends on growth objectives and the competition in the business. The company believed this level of investment was necessary to maintain its technological lead.

▼ *Return on Total Assets*
This ratio measures the productivity of the company's assets. It is calculated by dividing net income by total assets. The ROA for Trimble is calculated as follows:

Return on total assets = net income/total assets

$500/$18,607 = 2.7%

This is also a low number if compared to the return on 30-year treasury bills, which were yielding about 8.5% at the close of the 1980s with the full faith and credit of the U.S. government. Why would an investor risk his money in Trimble stock? *Growth!* Remember, this is high-tech and the promise of future riches with high price earnings multiples that outstrip the earning power of bonds by orders of magnitude drives these investments. A public company with an operating history of five to ten years would be expected to produce a 15% return. This would translate into an additional operating income requirement for Trimble of $3,600,000, which is achievable with a reduction in research and development of $5,000,000.

▼ *Return on Net Worth*
This ratio is sometimes referred to as the return on shareholder equity and is calculated for Trimble as follows:

Return on net worth = net income/shareholder equity

$500/$5607 = 8.9%

The return on net worth is significantly higher than the return on assets, due to the amount of leverage the company is employing, but it is still rather paltry by comparison to some of the high quality public companies. For example, the return on net worth and return on assets for both Hewlett-Packard and DEC were identical at about 15% in 1989. Both companies are noted for their conservative management philosophy and avoidance of debt on their balance sheet. However, as previously discussed, higher leverage makes higher returns for the stockholders, a definite plus for a closely held company about to go public. A return in the range of 15% would be typical of a good performing, public, high-technology electronics company in 1989.

In late 1990, Trimble received large orders for its military product for the Desert Storm operation in the Middle East. This was primarily due to their technological leadership which the high research and development investment provided. Profitability also increased dramatically. Sales projections for 1991 were for over $100,000,000! The high investment strategy seemed to be paying off.

Using the Financial Statements

What does all of this have to do with making electronics products? By understanding these simple financial measures of performance, you can compare the performance of your company to industry standards. This is what outside investors and bankers will do. They may not fully appreciate your unique product contribution or speak its unique language, but they will understand if your business is being properly managed. Appreciation for these simple financial measures will help you plan the necessary financing to grow your company.

High-tech manufacturing companies are very complex companies to build. The risk of product and market development is compounded by the financial commitments for inventory and equipment to manufacture a product. By paying attention to the warnings that can be gleaned from the financial statements, you can avoid serious cash flow problems. For example, the following questions can be answered by careful study of the financial statements:

▼ Is the inventory turn ratio at the proper level?

▼ Are you collecting your receivables in a timely manner?

▼ Is the return on assets employed a reasonable number, or are you buying too much equipment ?

▼ Do you have adequate liquidity (cash and short-term investments) to meet your payables and payroll commitments?

▼ Are your profit margins at the right level and on target?

▼ Are your expenses under control?

You still have to develop, manufacture, and market your products with the highest skill, but watching the financial indicators can help you avoid trouble.

Cash Is King

This is really the bottom line. A company only exists to generate cash—it's the only thing we can actually spend without liquidating assets or borrowing on assets. Positive cash flow is most critical, for without it a company can become insolvent and be pushed into bankruptcy. In the high development start-up phase, cash flow will obviously be negative; there isn't any sales revenue. When a company enters the production phase, the two most important questions to be answered are when will it reach profitability and when will cash flow become positive. Most entrepreneurs direct their attention toward designing, manufacturing, and selling the product. At the end of the month, they look at the P&L statement to see if they made any money; however, they should also be looking at the checkbook to see if there is any money in the bank to pay all the bills. By paying attention to the following areas that affect cash flow, you can avoid getting caught with an overdrawn checkbook. Let's examine the effect of the various components of the financial statements on cash balances.

▼ *P&L Statement*
The P&L is summarized by the operating income. If you made more than you spent, you generated a profit for the reporting period which ultimately becomes cash. This will be clear as we develop the cash flow statement.

▼ *Accounts Receivable*
As we discovered with Trimble Navigation, if we can reduce accounts receivable, we generate immediate cash.

▼ *Inventories*
If we increase inventories, we will consume cash. This is very risky in the high-tech business as Apple Computer found out in 1989 when they bought a lot of dynamic random-access memories (DRAMs) at inflated prices because of a fear of a parts shortage. The high prices significantly increased their cost of sales and inflated their inventory.

▼ *Property and Equipment*
Increasing this item consumes cash unless it is financed by debt.

▼ *Depreciation*
This is a source of cash. When an item is depreciated, there is no cash outlay, but the value of the asset is decreased. It is strictly an accounting item that gets added back to the cash calculation.

▼ *Notes to Bank*
This is a source of cash; the interest on the note is charged against income on the P&L and the payment on the principal is a charge against cash. The payment reduces the amount due on the note on the balance sheet.

▼ *Accounts Payable*
If payables are increased, they are a source of cash because you get the use of the materials and services without immediately paying for them.

▼ *Accrued Liabilities*

Any accrued liabilities, such as wages or taxes, are a source of cash until they come due.

An operating cash flow statement can be constructed with these items. It's convenient to define working capital:

working capital (WC) = accounts receivable (AR) + inventory (INV)
– accounts payable (AP)

The change in working capital is the difference between the value at the end of two successive reporting periods. It can then be shown that the cash flow during a reporting period, such as one month is given by the following equation:

$$Ce = Cb + I - \Delta WC$$

Where Ce = the cash at the end of the month, Cb = the cash at the beginning of the month, and I = the operating pretax after interest income for the month from the monthly P&L statement.

ΔWC = the change in working capital for the month

The change in working capital accounts for the increase or decrease in accounts payable, accounts receivable, and inventory for the reporting period. For example, if inventory and accounts payable remained the same and accounts receivable increased, then the working capital would have increased for the period. A projection must be made for the rate at which receivables will be collected, inventory will be purchased, and the company's bills will be paid to generate an accurate cash flow projection.

There are other items which also affect the cash flow projection such as depreciation, taxes due, and bank borrowings, but they can easily be handled as follows:

Depreciation—A noncash item, add monthly charge to ending cash

Loan principal payment—Subtract from the ending cash

Taxes due—Subtract from the ending cash

New stock sales—Add to the ending cash

Bank loan—Add to ending cash

Other balance sheet items can be treated in a similar fashion. The spreadsheet in Figure 10-1 shows a hypothetical calculation for the monthly cash flow for Trimble Navigation for fiscal year 1990. The following assumptions were made:

1. Accounts receivable are collected in 90 days.

2. Accounts payable are current at 30 days and grow at the shipments rate of 5% per month.

3. Company shipments are growing at 5% per month.

4. The company's operating income is 2% of sales.

5. The company is paying interest only on working capital debt.

6. The last reporting period was December 31, 1989. This projection covers the fiscal year ending December 31, 1990.

7. The company will not significantly increase its fixed asset base during the period. Depreciation is fixed at $114,000 per month.

8. The company will not issue any new debt or obtain new equity financing during the fiscal year.

9. The company must make quarterly estimated income tax payments for 1990 of $104,000 and a payment for 1989 of $167,000 at the end of the first quarter.

10. Inventories grow at 5% per month.

Some comments will help interpret this cash flow projection. First, getting started on a cash flow projection is always a problem. This technique requires that you know the working capital for the period immediately preceding the first period—in this case January—and the rate at which accounts receivable are collected—here within 90 days. Of course, you also have to model the other components of the cash flow statement, but these are not too difficult to estimate.

This cash flow statement shows that Trimble will consume over $5,000,000 in cash for the period. Most of the cash will be used to finance the rapid growth in working capital from $10,100,000 in the beginning to $14,900,000 at the end of the period. This is the price to be paid for rapid growth of 60% per year. Trimble would have to either borrow more money or raise additional equity financing to avoid insolvency to carry out this scenario of high investment in growth. In fact, the company completed a successful public stock offering in July 1990 and raised $30,000,000 of equity financing.

Hypothetical Cash Flow for Trimble, Accounts Receivable Collected in 60 Days, Operating Profit 2%

Twelve Month Cash Flow ($000)	Jan	Feb	Mar	Apr	May	Jun	Jul	Aug	Sep	Oct	Nov	Dec
Beginning Balances												
Period Beg Working Capital	10182											
Period Beg Cash	515											
Shipments	3417	3588	3767	3956	4153	4361	4579	4808	5048	5301	5566	5844
Working Capital												
Acounts Recv, Period End	6125	7005	7355	7723	8109	8514	8940	9387	9587	10349	10867	11410
Inventory, Period End	5501	5776	6065	6369	6687	7021	7372	7741	8128	8534	8961	9409
Accts Pay, Period End	3417	3588	3767	3956	4153	4361	4579	4808	5048	5301	5566	5844
Working Capital	8209	9193	9653	10136	10643	11175	11733	12320	12936	13583	14262	14975
Change in Working Capital	-1973	984	460	483	507	532	559	587	616	647	679	713
Net Profit	68	72	75	79	83	87	92	96	101	106	111	117
Depreciation	114	114	114	114	114	114	114	114	114	114	114	114
Loan Payments												
Capital Expenditures			271									
Taxes						104			104			104
Bank Loan												
Stock Sale												
Ending Cash	2670	1872	1330	1041	731	296	-57	-434	-939	-1365	-1819	-2405

Tax Requirements at FY End

Fiscal Year Earnings	1088
Tax Obligation at 40%	435
Tax Estimates Paid	416
Taxes Due	19
Cash Available	-2405
Cash Surplus (Deficit -)	-2424

Figure 10-1: A hypothetical cash flow statement based upon collecting receivables in 60 days and an operating profit of 2%.

Figure 10-2 shows the effect of increasing operating profit margins from 2% to 15% and improving collections from 90 days to 45 days. This scenario generates $7,100,000 in cash over the previous one. Improving collections by 45 days provides $2,900,000 of extra cash. However, there are reasons that Trimble may not be able to raise additional cash in this manner.

▼ The company may have to invest heavily in research and development to meet competition and build market share.

▼ Competitors may be offering customers very attractive payment terms that stretch out receivable collections.

▼ Competition may be high forcing low profitability.

It's important for you to understand the constraints and options at your disposal to avoid a cash crunch that can sink your business.

Growing the Business

One of the goals for most start-up companies is to grow the business for sale of the company or a public stock offering. This is certainly a goal for any outside investors. It not only takes new products and increased sales to grow a company, but also adequate financing for inventories, receivables, plant, and equipment. A company can finance growth in three ways: debt financing, equity financing through sale of stock, or by using internally generated funds. The first two methods are fairly obvious; you go to the bank or to prospective investors and convince them to loan or invest money in your company. Although these may be the only alternatives to finance a high growth rate, they can be expensive in the case of equity financing and risky if the debt load gets too large for the company.

The last method, using internally generated funds, is usually the cheapest, and most conservative method, but it requires careful management of the company's assets, expenses, and profits. It may also slow your growth. Let's examine how this works with the financial measures we have learned about in this chapter.

First, it's useful to develop an equation that relates profit percentage to growth rate. Sales growth requires growth in assets such as plant, equipment, inventories, and accounts receivables. These assets must be paid for out of after tax profit to internally finance the growth. The required profit margin can be calculated as follows:

Hypothetical Cash Flow for Trimble, Accounts Receivable Collected in 45 Days, Operating Profit 15%

Twelve Month Cash Flow ($000)	Jan	Feb	Mar	Apr	May	Jun	Jul	Aug	Sep	Oct	Nov	Dec
Beginning Balances												
Period Beg Working Capital	10182											
Period Beg Cash	515											
Shipments	3417	3588	3767	3956	4153	4361	4579	4808	5048	5301	5566	5844
Working Capital												
Acounts Recv, Period End	5000	5254	5516	5792	6082	6386	6705	7040	7392	7762	8150	8558
Inventory, Period End	5501	5776	6065	6369	6687	7021	7372	7741	8128	8534	8961	9409
Accts Pay, Period End	3417	3588	3767	3956	4153	4361	4579	4808	5048	5301	5566	5844
Working Capital	7084	7442	7814	8205	8615	9046	9498	9973	10472	10996	11545	12123
Change in Working Capital	-3098	358	372	391	410	431	452	475	499	524	550	577
Net Profit	513	538	565	593	623	654	687	721	757	795	835	877
Depreciation	114	114	114	114	114	114	114	114	114	114	114	114
Loan Payments												
Capital Expenditures			271									
Taxes						104			104			104
Bank Loan												
Stock Sale												
Ending Cash	4239	4534	4569	4886	5213	5446	5795	6155	6424	6809	7208	7518

Tax Requirements at FY End	
Fiscal Year Earnings	8158
Tax Obligation at 40%	3263
Tax Estimates Paid	416
Taxes Due	2847
Cash Available	7518
Cash Surplus (Deficit -)	4670

Figure 10-2: Reducing collection of receivables to 45 days and boosting operating income to 15% produces a dramatic improvement in cash flow position.

New assets = growth rate * total assets

Profit dollars > new assets

Profit dollars > growth rate * total assets

By dividing both sides of this equation by annual sales revenues we obtain:

(profit dollars/sales) > growth rate * (total assets/sales)

(total assets/sales) = 1/asset turnover

(profit dollars/sales) = profit percentage

profit percent > (growth rate/asset turnover)

or

The limit to internally-funded growth rate!

> | growth rate < (profit percentage * asset turnover) |

At some point in time, the company must make enough profit and turn over its assets at a high enough rate to finance its growth, or reduce its growth rate. Investors and lenders will not continue to pour money into a company forever to finance its growth. Let's see how Trimble Navigation could grow from retained earnings.

1989 after tax profit = 1.6%

1989 total asset turnover = 1.7

Maximum internally financed growth rate = 1.6%*1.7 = 2.7%.

Clearly, Trimble has to borrow heavily or sell equity in the company to finance its high growth rate. Suppose that accounts receivable could be reduced to 60 days to increase the asset turnover and that research and development was cut to 15% of sales.

1989 after tax profit = 9%

1989 total asset turnover = 2.1

Maximum internally financed growth rate = 9%*2.1 = 18.9%

This is a significant increase, but still inadequate to finance the 50+% annual growth rate of the company. Trimble made a successful public offering of its stock in July, 1990 to raise $30,000,000 to finance its growth and retire debt. You should understand the financial strains high growth will place on your company and make appropriate financing arrangements to accommodate the growth.

The asset turnover method is a quick calculation to determine a company's ability to internally finance its growth. It assumes that all assets have to increase to meet the growth requirements, including plant and equipment. A detailed cash flow statement that includes working capital projections and plant and equipment investments will provide a more precise picture of the situation.

Monitoring Performance

The performance of your company cannot be measured without developing reports that provide the right information to your management team. The following reports can form the core of the management information system to allow you to evaluate the company's performance:

Monthly Reports

- ▼ Cash flow statement
- ▼ Profit and loss
- ▼ Balance sheet
- ▼ Inventory projection
- ▼ Sales forecast
- ▼ Shipment forecast

Weekly Reports

- ▼ Shipments to date
- ▼ Bookings to date
- ▼ Cash report
- ▼ Accounts receivable aging
- ▼ Accounts payable aging
- ▼ Purchasing commitments

The review of some of these reports can certainly be delegated to appropriate members of the management team, but it is important you review them as well. As Harry Truman said, "The buck stops here!" And in your company, you're the "here." The reporting mechanisms should be in place to generate all of these reports when the company begins to ship products. Prior to product shipment, a cash flow state-ment, balance sheet, and the appropriate research and development expense reports and schedules are adequate to manage the product development phase.

Garbage In, Garbage Out!

You have just received your end of the month financial reports and discovered that you missed your shipment targets and inventory is continuing to increase beyond projections for the third month in a row. You call a meeting with your manufacturing manager to review the situation. This person knows that shipments were missed again, but a critical IC was not available. You ask, "With all that inventory, why don't we have the right parts?" The reply? "The computer said we had it, but when we went to pull the kit, it wasn't in the box. We're looking into it."

This conversation takes place more often than most high-tech manufacturers would care to admit. Even with the best computer technology, companies have difficulty keeping track of all the parts in most high-tech products. Information-handling technology is not the source of this problem. The evolution of the computer information system has been marked by a remarkable abdication of the human thought process in managing the technology. How many times have you had an argument with a clerk in a large organization, such as a bank or an insurance company, over an entry on your statement only to be told "it's in the system and there's nothing I can do about it?" A manufacturing consultant remarked, "I'm amazed at how many computer terminal operators assume that if it's in the computer it's right. Many seem to exercise no independent thought at all!"

What these people—and often their managers—don't appreciate is that before information is available on a computer screen some human being had to key the data into the database from a piece of paper. This is the Achilles heel of modern information systems. A comprehensive paper flow system with adequate checks and balances must be in place before a computer information system can be of any use to an organization. Many entrepreneurs, particularly in the high-tech industry, believe that they can leave the paperwork behind at the large company they previously worked for. Nothing could be further from the truth, although the process may be more streamlined. How can you maintain control of your company and not strangle it with paperwork?

It's a difficult challenge, but one that can be met by enlisting the services of your own staff including an experienced accountant familiar with manufacturing, or, alternatively, an outside consultant that specializes in setting up internal control systems for manufacturing companies. The process is to walk through every transaction that is made within your company and determine the necessary paperwork required to provide an audit trail and develop a procedure for handling the paperwork. If you already have people on board doing these tasks, involve them in the process of defining the procedures and the paperwork. Otherwise, set it up with the manager and outside expert. The procedures should be documented and placed in a manual that will be used as a reference for the company and to train new employees.

There are many areas that need to be addressed in a high-tech manufacturing company. The following partial list illustrates the principles involved in this process:

▼ *Materials*

Purchase order procedure that covers how material requisitions are handled and purchasing limit authority.

Receiving procedure that covers the disposition of material such as inventory and capital equipment, and quality assurance requirements.

Inventory counting and audit procedures including material tags, sign-out sheets, and storeroom controls.

▼ *Manufacturing*

Job work order procedure that describes how work orders are opened or closed and labor and material costs are reported.

Returned goods procedure that describes how product returns are received and processed.

Month end closing procedure that describes the reports and audits to be performed before the month end is closed.

▼ *Sales*

Sales order processing procedure that describes how new sales orders are booked by the company.

Demonstration equipment procedure that describes how demo equipment is tracked.

▼ *Accounting*

Accounts payable procedure that describes the process and the necessary paper work required for bills to be paid, such as pur-chase orders, invoices, and material receivers.

Accounts receivable procedure that describes the processing of checks for payments on account and the deposit of the checks to the bank.

Payroll procedure that describes how payroll is processed and allocated to jobs.

The paper control system should be set up to interface with the computer information system that you chose and should tie into your books. If it is done independently, you may have to modify the software you purchased to work with your internal control system. Software modifications are an expensive and time consuming task, and are often not supported in standard software updates.

Can you keep your company from becoming bloated on paperwork? Yes, but it will take discipline and use of the latest technology. One company president we know of had a policy in place that successfully violated the third law of paper thermodynamics. After a workable paper tracking system had been put in place, he placed a copy of every form in use at the company on a wall. The advocate of a new form had to "go to the wall" and replace an existing form.

Minding the Store

As much as we would like to assume that all of our employees are hardworking and honest, it just isn't so. From time to time, we read about the trusted bookkeeper that embezzles $750,000 or shipping clerks that ship thousands of dollars of product to their home before being discovered. Our immediate reaction is "that couldn't happen to me," but it can. A well-defined paper control system and segregation of duties can eliminate this risk.

It's relatively simple to implement. Review your procedures and determine if one person has total control over any process that could involve diversion of funds or material. Your accountant or bank can help you with a detailed analysis of your operation and suggest ways to minimize your exposure as illustrated in these examples.

▼ The person that opens the mail and logs checks received should not also make accounts receivable entries.

▼ Bank statements should be reviewed by someone other than the bookkeeper.

▼ The person that signs the checks should be different than the person that prepares the checks.

▼ The person who issues purchase orders shouldn't be authorized to approve receipt of goods.

Segregation of duties can help you avoid unpleasant surprises and is one of the important policies you should have in place.

Automating the Process

It is doubtful that computer systems will ever eliminate the need for paper reports because of the requirement of audit trails for financial transactions. It is too easy, even in highly secure systems, to modify the database. However, a good computer system is invaluable in providing high efficiency and timely information to manufacturing companies in the 1990s. The technology is affordable for even the smallest of manufacturers. If your plans call for a multi-product company, or for sales greater than $500,000, you will want to evaluate software packages that provide manufacturing resource planning (MRP) integrated with general ledger.

General ledger provides all of the accounting functions, such as accounts receivable and payable, profit and loss statements, and balance sheets. MRP provides all of the information to manage the manufacturing process and is integrated with sales order processing. Modern systems are all transaction based so that the users of the system control the data that is entered into the system. Data is available on a near real time basis for use by management and staff for decision making.

A successful implementation of a computer information system requires top management commitment, involvement of the staff that will be using the system, excellent training of the staff, careful selection of the system, good system documentation, and support from the vendor. The system should be integrated so that all software modules are interconnected and share data from all parts of the organization. For example, manufacturing data such as labor costs should be available to the accounting department for payroll and job cost analysis.

There were over 300 MRP software alternatives available to the new high-tech company at the end of the 1980s. The least expensive way to get started with the most capability and minimum initial investment is to use a timeshare service offered by some of the larger vendors. For example, both Compaq and Trimble Navigation started by using ASK, Inc.'s service to implement their management information system. As the companies grew, they installed their own system using ASK's software. Since the timeshare service capability was identical, disruption during the changeover to an internal system was minimized.

There are also many programs available for operation on personal computers; the better products are multi-user and can accommodate a fairly large number of parts in the inventory master. Almost all of the products of any consequence satisfy the accounting requirements, but the manufacturing resource planning capability may be severely limited on many of the offerings. Study this part of the product carefully to be sure that it will meet the needs of your company.

The next grade of capability is software that runs on minicomputers or networked personal computers. These systems are very sophisticated and probably suitable for companies that have sales from $1,000,000 to over $1,000,000,000 annually. They can handle all aspects of a manufacturing company from materials control to human resources planning.

The large array of products that are available makes it imperative that you understand and define your needs before you embark on a shopping trip. Before you start production, your management team should develop a specification for requirements that addresses each department's needs. If you lack the technical expertise on your staff to develop the specification and evaluate the available alternatives, don't hesitate to bring in an outside consultant, either an independent or someone from your accounting firm. Be sure that whoever you use has the necessary operating and educational background in manufacturing and finance to do the job. Practical experience is invaluable in this process, because you will be setting up the complete control system for your company. Mistakes in the selection of the software vendor and the development of the paper control system can be much more expensive than the price to be paid for expert advice.

Once you have found a software and hardware package that meets your projected needs, insist on getting references, carefully review the documentation, and evaluate the company's support package. The support in training, software upgrades, and bug fixes is extremely important to the success of installing an automated management information system. Regardless of your system choice, remember that computers are governed by GIGO—garbage in, garbage out.

CHAPTER 11

How to Get Started

*P*robably the most common question that entrepreneurs ask goes like this: "There are so many things to be done; what should I do first?" In this chapter, we offer one answer to that question.

Here are 18 issues that have to be addressed, one way or the other, at the outset of your venture. They are listed in approximate chronological order of the "typical" start-up experience. We have chosen to terminate the list at the point where the design team is beginning the final product design and the marketing team is beginning to plan the product's market introduction.

Clearly, every start-up company will have different needs, so this list is not entirely correct for every company. But we hope it will serve as a guide in deciding what to do, and in what sequence, when you first take the plunge.

1) *Gather the founding team of people.*

   There is no magic formula for how many people should be on a founding team, or what type of people they should be. The requirements for this group are mainly that each person has a genuine contribution to make and that everyone is prepared to make a real commitment (that is, sacrifice) to the enterprise.

2) *Develop the product idea, and the business idea wrapped around it, in some detail.*

   The founding team needs to spend some time together hashing out the business idea. Start by discussing technology, product design, and patents. Where are the innovations that you hope to bring to market? Exactly what products are you planning to build? What technical expertise will be required in your product development group? Just what is your uniqueness?

   Next, identify the customers for your product. Spend some time on refining your market niche—the more carefully defined, the better. Now climb inside your customer's head and try to understand the competitive advantage of your new product. Put together a preliminary positioning statement.

Talk about how the product will be manufactured. Your ability to manufacture is a cornerstone (along with product development and marketing/sales) of your new company. You cannot put this discussion off. It must be part of your initial planning.

Likewise, you must plan, as best you can at this stage, your product's distribution. The main reason to talk about sales is not so much as to arrive at answers as to just identify the questions. There will be opportunities to discuss sales organizations as you progress, so you can refine your ideas as you go along.

It is a good idea to discuss among the founders the eventual size and organization of the company. These discussions will help you to decide upon the initial organization.

3) *Do the necessary market research.*

When you visit potential customers, it will be easy to discuss specifications and softer performance parameters with them. But remember that you seek other information as well, such as inputs on distribution channels and post-sales support. As you talk with these important people, keep in mind the fundamental questions for your enterprise: Is this business idea feasible? Is there genuine demand for this product? If so, how much?

4) *Assign who will do what.*

Having decided to move ahead with your company, it's time to allocate the work of the business among the founding team. It is likely that your team is incomplete, and is missing one or more key players. It is best to identify at this time the individuals you would like to recruit into those jobs.

This is a serious discussion. You need to ask yourselves if the founding team contains the right people for the key positions in the company, including the president's position. It's not uncommon for the team to acknowledge they lack a president and for them to recruit one. And it is quite common for funding sources, especially venture capitalists, to assist in the recruitment.

5) *Decide upon a name.*

It may surprise you to see this activity so early on this list. But we have found that it's advantageous to have a name for your company when it's time to consult with an attorney or other resources, and especially when you begin the fund-raising process. A fledgling company with a name carries much more credibility than just a group of people.

6) *Hire an attorney.*

The services of an attorney are essential when planning the formation of a new company. There are many legal details involved in founding a company, and you need sound advice to insure you're doing it correctly. Issues such as the legal form of the company, classes of stock, and ownership of the founders are not to be toyed with. You should be able to find a law firm experienced in start-ups without too much trouble.

7) <u>*Write the business plan.*</u>

No matter how you plan to finance the company, you should prepare your business plan early in the start-up process.

The preparation of the plan is a lot more than just writing—that's the easy part. You need to have thorough discussions among the founding team on an entire sequence of subjects:

▼ What are our objectives?

▼ What strategies will we implement in pursuit of those objectives?

▼ What tactics will we use to accomplish our objectives?

Having answered these questions, and having identified your uniqueness and consequent competitive advantage, you can begin writing the plan.

Here is an appropriate time to <u>discuss the culture you wish to create in your new company</u>. You'll find that culture is an issue as large as the other important issues you're discussing, and that it relates to some of your business strategies.

8) *Secure your seed funding.*

A major decision to be made as your prepare the business plan is how the company is to be funded. For many start-ups, it's appropriate to start with modest seed funding to get the company through the product feasibility phase or some later phase such as the completion of test marketing. For others, it's best to secure major funding required for product development and distribution channel development right away. And in some situations, the founders will contribute all the seed funding (and work without a salary for a period of time). In any case, here is the time to provide for the initial funding of your company.

9) *Quit your jobs.*

Somewhere along the line, your start-up company has to mature from an evenings and weekends activity to full-time. Ideally, you shouldn't have to quit your job until the required seed funding has been committed. But many entrepreneurs find that a full-time commitment is needed before then. Once the seed funding is there, however, you and the other founders should be fully committed to the enterprise to give it the energy it needs.

10) *Identify your board of directors.*

Now that your company has some life of its own, you can interest highly qualified people to serve on the board of directors. It may be best to identify some members of the board earlier than this if you plan to seek substantial seed funding, since sources of funding are often influenced by your choices for the board.

11) *Incorporate.*

This step in the process is relatively simple providing you have good legal counsel. The many decisions to be included in a founders' agreement can be worked out readily with expert assistance.

12) *Hire an accounting firm.*

As soon as you begin to keep financial records, it is best to hire the services of an accounting firm. They can help with a variety of tasks, such as corporate reporting requirements, tax filings, the establishment of your accounting systems, and more.

13) *Lease and equip a facility.*

Once seed funding is secured, your company should have a facility. Customers, vendors, and potential business partners will be far more impressed if they can visit you at a facility rather than somebody's garage. Also, you will find that working together away from home promotes unity among the team.

A word of advice: your facility should be functional, but not necessarily attractive. Don't waste your money on large offices, expensive furniture, and rented houseplants.

Many aspects of a start-up can be funded on a shoestring, but you shouldn't scrimp on test equipment or other gear directly related to the quality of your product. Of course, for some equipment it may make sense to lease rather than buy.

14) *Determine the feasibility of the product design.*

Here is a major checkpoint. You should strive to validate your product idea before you have committed much money to the enterprise. There have been some spectacular failures of high-tech start-ups that violated this rule; don't join that club! You owe it to your stockholders, your partners, your family, and yourself to walk away from an idea that's not viable.

15) *Secure the first round of funding.*

Your start-up company is now moving into a phase in which larger sums of money will be spent for parts, equipment, tooling, salaries, market introduction, initial inventory purchases, and much more. It is likely that additional funding beyond the seed amount will be required to get the company through the prod-uct launch and initial shipments. If your seed funding was modest, here is the time to update your business plan and raise needed funds.

16) *File patents.*

Of course, as soon as feasibility of your ideas is demonstrated, you should file for patents. Since filing does involve considerable effort and some monetary expense, it is wise to wait until this point to file.

17) *Prepare a detailed product definition.*

Having completed feasibility testing, you are now in a position to determine the product you wish to build. Establishing and maintaining contact with a number of potential customers will pay off handsomely here. You may choose to use alpha site testing of a prototype to help refine the product definition.

18) *Begin final product design.*

Now your start-up company begins its work in earnest. The next challenges will be to build marketing, sales, and manufacturing programs. Best of luck!

an Outline

1. ... ive Summary

1.1 ...mmary of business idea

1.2 ...embers of management team

1.3 ...ief description of products, current and future

1.4 ...ief discussion of markets

1.5 ...ief discussion of five-year financial projections

1.6 ...ief discussion of funding requirements

1.7 ...ef discussion of exit plan

2. ...s History (use this only if your company has begun ...ns)

2.1 Founders and founding date

2.2 Form of organization (corporation, research and development partnership, etc.)

2.3 Ownership (who owns how much stock)

2.4 Financial obligations (stock options, leases, etc.)

2.5 Summary of financial performance to date

3. Business Description

3.1 Mission statement

3.2 Discussion of uniqueness and competitive advantage

3.3 Objectives

4. Market Description

   4.1   Customer identification (be as specific as you can)

   4.2   Customer needs

   4.3   Market size, growth, and segmentation

   4.4   Distribution channels and purchasing influences

   4.5   Government regulation (requirements and restrictions on design, sale, or export)

5. Competition

   5.1   Competing products

   5.2   Competing companies

   5.3   Alternative techniques

6. Product Plan

   6.1   Technology strategy

   6.2   Products

         A. Key specifications

         B. Order of introduction

   6.3   Development plan (what, when, who)

7. Marketing Plan

   7.1   Product positioning and marketing strategy

   7.2   Pricing

   7.3   Sales forecast and market share

   7.4   Promotion

   7.5   Domestic and international sales organization

   7.6   Product support

   7.7   Compliance with government regulatory requirements

8. Manufacturing Plan

   8.1   Manufacturing strategy

   8.2   Organization

8.3   Facilities

8.4   Capital equipment

9.   Management Team

   9.1   Brief resumes

      A.   Key employees

      B.   Board of directors

   9.2   Descriptions of key unfilled positions

   9.3   Proposed salaries

   9.4   Proposed stock ownership

   9.5   Employment agreements (if any)

10.   Elements Critical to Success

   10.1   Market risks

   10.2   Technical issues

   10.3   Other business issues

11.   Exit Scenarios

   11.1   Preferred exit plan

   11.2   Other possibilities

12.   Financial Plan

   12.1   Past financial results (if any)

      A.   Income statement

      B.   Balance sheet

      C.   Statement of cash flow

   12.2   Five-year projection (Monthly first year, quarterly second year, then annually)

      A.   Income statement

      B.   Balance sheet

      C.   Statement of cash flow

   12.3   Assumptions used in preparing projections (Receivables, bad debt, interest payments, taxes, etc.)

13. Exhibits

    13.1 Resumes of key employees (essential)

    13.2 External analyses used in preparing the plan (optional)

    13.3 Product brochures (if company has begun operations)

14. Financing Package (if this plan will be used to raise funds)

    14.1 Funding requested

    14.2 Capitalization scenario (who will own how much of what type of stock after funding is complete)

    14.3 Reporting (how often investors will be informed of progress)

    14.4 Use of proceeds

The Product Proposal

1. Product Description

   1.1 Features, advantages, and benefits. Customers buy products because of the benefits to be gained. Besides a thorough description of the product's performance features, this section should outline the train of thought that the customer will follow from features through advantages to benefits.

   1.2 Relation of the product to others in your line. This constitutes another perspective on the new product. In some cases, this relationship is the key strategic attribute of the product. For instance, a new accessory could be conceived solely to enhance the sales of a mainframe product.

   1.3 Technology discussion. A product proposal is the intersection of new ideas about the marketplace and the technologies available to the company, including manufacturing technologies. The technological side of the equation merits lots of discussion at the start of a proposal.

   1.4 Uniqueness. The product's uniqueness, stated from the viewpoint of the potential user, is its competitive advantage. This section of the proposal describes the basic strategy behind the product idea and the contribution the product will make to the world.

   1.5 Applications. In order to fully understand the product, one must learn who will use it and what it will be used for.

2. The Product's Fit with the Company

   2.1 Mission statement. This is the first and most important test of the product's usefulness to your company.

2.2   Product strategy. It may seem obvious, but you must carefully evaluate the position of a proposed product in your line to see if it creates fundamental shifts in your planned product offering. Too often, managers of young companies miss this point and introduce follow-on products that change their whole plan. Another mistake is creating new products that are too similar to others, resulting in insufficient differentiation between products.

2.3   Related capabilities of the company. It's important to examine how a new product might require capabilities from your company that you do not yet have. You need to consider technological capabilities in product development, manufacturing, and product support. You also need to evaluate whether the distribution channels you now use will support this new product.

2.4   Company's history with related products. Much can be learned by examining what your company has already learned about markets, technologies, and competitors from its prior experience. Sales results and competitive situations of other products will often reveal your company's strengths and weaknesses—rich information to help forge strategies for new products.

3. The Product's Fit with the Market

3.1   Market analysis. The marketing member of the product definition team should include his or her market analysis in the product proposal. The analysis should include the following items:

▼   market size and growth

▼   market accessibility, including geography, distribution channels, entry barriers, exit barriers, governmental obstacles, and customer attitudes, motivations, and habits.

▼   market durability, including competing technologies, trends changing the market from within, and external change agents.

▼   market structure, including customer segmentation, identity of decision-makers, distribution channels, and competition.

▼   technical needs of customers.

3.2   Competitive analysis. Your competitive advantage can be judged only in comparison with other products and companies. So a competitive analysis is essential to understanding any new product's potential. This analysis should include not only data on competing products, but also on the companies that make them—technological capabilities, manufacturing strengths, distribution channels, and financial backing. A thorough understanding of your competitors' strengths and weaknesses can help you considerably in preparing your strategies.

3.3   Positioning. With the above data on markets and competitors, you have all the information you need to develop a positioning statement for the new product and the promotional strategies that flow from it.

3.4   Pricing and profitability. With knowledge of market conditions, the competitive situation, and your internal cost estimates, you are prepared to develop a pricing scheme for the product.

3.5   Sales forecast. Only after you have arrived at the price can you make a good estimate of sales. The product proposal should include a rationale for the forecast, which will usually include a graph of the volume-price relationship. Accompanying the price should be a statement of the expected profitability of the new product.

4.   Return on Investment

4.1   The first step in analyzing ROI is to estimate the costs to develop the product. These costs include research, product development, capital equipment purchases (including tooling), and market development costs (promotion, sales training, etc.). Since an ROI calculation uses cash transactions, use both capital and expense items.

4.2   The next step in ROI is to identify the return from the product, derived from your estimates of sales and profitability.

4.3   Finally, one can calculate ROI using any of several techniques from simple ratios ("return factors," etc.) to more sophisticated discounted cash flow calculations of internal rate of return (IRR), which are easily available on PC spreadsheets and financial calculators. We recommend the comprehensive IRR calculation.

5. Sales and Support

5.1 International considerations. Some aspects of a new product may have considerable international repercussions. For example, a product which cannot run on 50 Hz power would need special effort to sell outside the United States and Canada, or perhaps should not be exported at all. And some products, of course, will be ideal for international markets, such as a troposcatter telecommunications system for developing countries.

5.2 Distribution plan. The selection of a distribution channel is so important that it must be considered at the product definition stage.

5.3 Product support plan. Likewise, support is an issue that can make or break a new product decision. A service policy and general description of how service will be provided are essential elements of a product proposal.

5.4 Timing considerations. How many times have you heard of a product, company, or executive who was "in the right place at the right time"? Those situations were not always luck. You should consider any market factors which have timing implications for your new product. Sometimes a new product needs to be launched with haste, or held back a few months. One of the greatest marketing sins is to introduce a product just as the market "window of opportunity" is closing.

6. Manufacturing Plan

As with product support, it's unlikely that you will have a complete picture of how the new product will be manufactured. But an outline of the manufacturing plan is important at the definition stage, especially if elements of competitive advantage are involved.

Volume–Price Graph

The purpose of a volume-price graph is to quantify your knowledge of competitive products and market dynamics. The resulting curve represents a forecast of your product's sales as a function of its price.

There are four items of information you need to have before constructing your volume-price curve. These data are gathered as part of your market research and competitive analysis. They are:

1. Market size, with emphasis on the total potential customer population.

2. Competitors' prices and sales volume.

3. Customer perception of the value of your product compared to competitors' products.

4. Alternative techniques—their cost to the customer, frequency of use, and value to the customer compared to your product.

We will illustrate the technique by walking through the creation of a hypothetical volume-price graph, as shown in our diagram. First, draw a horizontal axis representing price and a vertical axis representing units sold per unit time. Locate each competitor at their price position on the x-axis, and mark your estimate of their sales above that position. In our example, there are three competitors.

In some cases, competing products may be items constructed by the end-user rather than commercial products. This is common in the scientific research market, for example, where graduate students often construct their own apparatus since they lack funding to buy it. In the example, we have located the position of such homemade alternatives on the x-axis and estimated units constructed per unit time with another mark. Now we can begin to study market behavior.

Let's assume that our hypothetical product is viewed by customers as superior to those of competitors A and B but inferior to that of industry leader C. We can now analyze the orders we would receive at various price points. For the example, we consider ten points, starting with the highest price imaginable, a price equal to competitor C:

1. If we set our price equal to C, we won't sell anything. This price is too high to be able to steal orders from B, and C's customers have no incentive to switch.

2. If we set our price halfway between B and C, we will steal some orders from C and some from B.

3. If we set our price equal to B, we will steal many of B's orders. (Only those customers loyal to B will pay the same for B's inferior product.) We will also win many of C's orders and even some of A's orders. Our sales volume will be about equal to the original volumes of A and C.

4. If we set our price halfway between B and A, we will gain even more orders from B, and the last available to us from C. (If C's customers haven't switched by now, they never will.) We will also gain significant orders from A, now that our price is so close.

5. If we set our price equal to A, we will clean up the remaining available orders from B and steal many more from A. Only A's most loyal customers will not switch at this point.

6. With a price somewhat less than A, we will pick up a few more A customers and start to convert a few users of homemade alternatives. But the price is too much above the homemade items to allow a heavy penetration of that segment of the market.

7. As the price gets closer to the homemade alternatives, we convert many of them to our customers.

8. With our price set equal to that of homemade alternatives, we capture a great deal of that business.

9. If we set our price lower than the homemade alternatives, we capture the vast majority of them and begin to lose any further growth potential.

10. Eventually, any volume-price curve reaches a plateau at the low end of the price scale. This is where the price is so low that everybody who needs one will eventually buy one. Assuming that the product in question is not a consumer item, the plateau will be in the same order of magnitude as the rest of the curve.

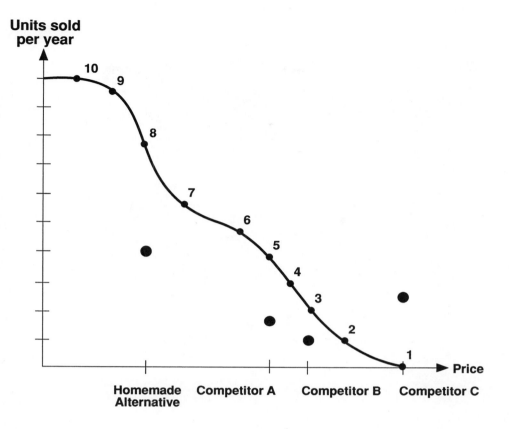

The New Product Marketing Plan

Many of the entries in this outline are identical to entries in Appendix B, "The New Product Proposal." Rather than repeat these descriptions, we have denoted the entries with an asterisk (*) so that the reader can refer to descriptions in Appendix B.

1. Product Description

    1.1 Features, advantages, and benefits.*

    1.2 Uniqueness.*

    1.3 Relation of the product to others in your line. This is the place to briefly review the product's position in your overall product strategy. The reader of your plan must understand whether the new product will obsolete any products in your current line, complement them, or clash with them.

2. Market Situation

    2.1 Customer description. You should be able to identify who the customer is to some degree of detail. Industry, geographical location, department, and job title are all elements of such a description.

    2.2 Applications for the product.*

    2.3 Market analysis.*

    2.4 Competition. For each competitor, you should be able to describe a number of attributes:

    A. Strengths and weaknesses, compared to your company.

    B. Positioning in the marketplace.

    C. Share of market.

      D. Coverage of various market segments.

      E. Geographical coverage.

      F. Trends in its business that might affect you.

3. Company Situation

    3.1 Sales history of related products.*

    3.2 Marketing resources. There will be specific resources your company needs to be able to market the product. It is especially important to identify those resources that you do not yet possess and to discuss how they will be acquired.

4. Product Objectives

    4.1 Sales and market share, by segment. It's not enough just to set objectives for sales levels and market share. You should also identify the niches in which you will play and examine how well you can compete in them. Successful businesses tend to dominate the niches they occupy.

    4.2 Profitability. The targeted profitability of a new product is a parameter of great interest to all on the company management team. Any new product that does not meet the company's targeted average profitability should be viewed critically. (Never deliberately introduce a below-average product. Too many will happen anyway despite your best efforts!)

    4.3 Timing of the introduction. The plan should describe the latest product development schedule and the manufacturing schedule for the product at hand. These schedules should be compared to your assessment of market "windows of opportunity" that exist.

5. Marketing Strategy

    5.1 Positioning statement.

    5.2 Penetration strategy by market segment. Here is the key to your introduction strategy. How will you penetrate the various segments of the market? For one, you might concentrate on taking a few key accounts away from competitors. For another, you might depend on a particular strategic relationship with another company. There are many, many possible strategies.

5.3 Anticipated competitive moves. This subject must be treated somewhere in the plan. You may want to make it part of the above market penetration discussion, the pricing discussion, the competitive analysis, or put it here.

5.4 Pricing strategy. All elements of marketing strategy tie together. Here you should describe how the pricing supports the strategies for market segment penetration. The numerical analysis is covered in a later section of the plan.

5.5 Distribution strategy. A short discussion of distribution plans and their tie to market penetration should be included in the strategy section.

5.6 Post-sales support strategy. Likewise, the general approach to support should be described here. Will support people be located at the factory only, or also in the distribution outlets? Will the service approach be board exchange or component-level repair?

6. Promotion Plan

6.1 Literature. Your literature plan should include both items aimed at the customer—data sheets, brochures, configuration guides, applications notes—and items for internal consumption, such as competitive analyses.

6.2 Publicity. The best form of promotion is for the magazines and newspapers read by your customers to contain favorable articles about your product.

6.3 Direct customer contact. A new product message is powerful when delivered directly to the customer. This can be accomplished with seminars, demonstrations, customer visits, and exhibits at trade shows.

6.4 Indirect customer contact. Advertising, direct mail, and their variations are means to reach a larger audience.

7. Timetable

7.1 Product availability.

7.2 First public announcement.

7.3 Literature.

7.4 Training.

7.5 Promotion.

8. Distribution Channels

    8.1 Domestic. It is imperative that your domestic channels be identified, placed under contract, and trained before the first product is introduced. There is nothing more deadly to a start-up company than customers not being able to buy your product when they hear about it.

    8.2 International. Even if you don't plan sales to international markets right away, you should have an international distribution scheme thought out by the time this introduction plan is prepared.

9. Other Detailed Plans

    9.1 Trial marketing. If you are engaged in trial marketing of the product, the program should be described in detail here. The coordination of trial marketing with other introduction activities can be crucial.

    9.2 Packaging. For consumer products, packaging is an important element of the marketing plan. The subject may deserve considerable discussion here.

    9.3 Sales training. Occasionally, training of the people in the distribution channel will be complicated enough to require a description in your marketing plan.

    9.4 Post-sales support. It is a good idea for the person in charge of post-sales support of the product to write this portion of the marketing plan. Topics should include repair strategy, customer telephone support, spare parts distribution, and training of support personnel. In the case of a complex product or organization, this may require a separate plan.

10. Financial Analysis

    10.1 Pricing and profitability. Here is where you should explain the methodology used to arrive at the product's price. And you should discuss fully the resulting profitability.

    10.2 Investments in the product. All investments in the project leading to the product introduction should be listed in order to be able to evaluate your company's return on investment. The costs of research, product development, manufacturing tooling and other equipment, the marketing introduction, and preparation for customer support are all investments which should be included.

10.3 Sales forecast. In this section you need to list expected sales of the product over its lifetime, typically five to ten years. It is often helpful to break sales down into categories—for example, dividing sales by geography can illustrate the impact of a gradual introduction of the product into foreign countries.

10.4 Return-on-investment (ROI) calculation. Given the information in the above three paragraphs, you can calculate your company's return on the funds invested into this new product. The preferred form of ROI calculation is internal rate of return (IRR), which is available as a function on most PC spreadsheets.

Sales Training Seminar

1. Product

   1.1 General description and nomenclature.

   1.2 Capabilities and specifications.

   1.3 Demonstration by the instructor.

   1.4 Relationship to other products, including those from the same company.

2. Applications

   2.1 Uses of the product.

   2.2 Key customer needs, and their relation to product performance.

3. Customers

   3.1 Identity: Industry, function, title.

   3.2 Sophistication (for example, level of computer literacy).

   3.3 Mentality (conservative or risk-taking, etc.).

4. Competition

   4.1 Thorough comparison with directly competitive products.

   4.2 Competitors and their distribution channels.

   4.3 Forms of indirect competition.

   4.4 Strong points against the competition.

   4.5 Weaknesses against the competition (important!).

5. External Influences (such as government regulations or industry standards)

6. Positioning

   6.1 Explain the product's positioning statement.

   6.2 Approaches to customers.

7. Sales Aids (such as hardware demo kits, flip charts, software presentations, or videotapes)

8. Promotion

   8.1 Overview of promotional plans.

   8.2 Literature and videotapes.

   8.3 Publicity.

   8.4 Other activities, like trade shows.

9. Hands-On Familiarization with the Product (if feasible)

10. Review and Questions

Note: Some companies have successfully used role-playing or similar training techniques in sales seminars. A recent innovation is to practice your sales presentation in front of a video camera and critique yourself.

How to Plan for International Sales

The following outline[1] describes a plan that should be developed to focus your international sales efforts.

Introduction: Why the Company Should Export and Its Overall Export Goals.

1. Export Policy Commitment Statement From Top Management

2. The Situation/Background Analysis

    2.1 Description of products.

    2.2 Description of company's operations.

    2.3 Personnel and export organization.

    2.4 Internal resources of the firm.

    2.5 General background of the industry—its structure, competition, and demand for products and services.

3. The Marketing Component

    3.1 Market projections.

    3.2 Competitive environment.

    3.3 Identification, evaluation, and selection of target markets.

    3.4 Product/service selection and pricing objectives.

    3.5 Methods of distribution.

    3.6 Terms and conditions of sale.

    3.7 Internal organization (credit, administrative policy/procedures, insurance, and management reports).

    3.8 Sales and profit goals/ forecasts.

4. Product Support

    4.1   Sales volume.

    4.2   Training requirements.

    4.3   Documentation.

    4.4   Special country features.

5. Strategy for Penetrating Foreign Markets

    5.1   Countries where company has competitive advantages.

    5.2   Primary target countries.

    5.3   Secondary target countries.

    5.4   Indirect marketing efforts.

6. The Export Budget

    6.1   Pro forma financial statements—projected sales and expenses.

7. An Implementation Schedule

    7.1   Identify key activities and assign responsibilities

    7.2   Conduct periodic managerial review.

[1] This outline is based upon one developed by the U.S. Small Business Administration.

Product Development Program Manual

1. Market Research
   Adequate market research should be done before any new product development program is started. Investigations into new technologies, which can range from components to block diagrams, may be started in parallel or before market studies have been completed.

2. Product Definition
   The evolution of a new product definition usually involves a combination of market research or insight, technical expertise, and a lot of discussion. Many great new products come about because of an engineer's tinkering in the lab. Compaq was supposedly founded with a design scribbled on a restaurant napkin, so the process cannot be reduced to a regimented set of rules that can be set down in an expert system and run on a computer. Good products come about because of someone's insight into the way things can be.

3. Product Development Cycle
   The development cycle is intended to get a product to market in the shortest possible time and in a cost effective manner. Timely completion of the project should not be dependent on major scientific or engineering breakthroughs. This development should be undertaken in an investigation where there are fewer time pressures and the "burn rate" on funds can be more easily controlled. For example, if the new product depends on the development of an esoteric component or process to be successful, a lot of money can be at risk implementing all of the other features of the product such as

packaging, power supplies, and more common technology subassemblies if the new component cannot be produced.

3.1 Project Kickoff

A project business plan should be developed to initiate a new project. It should include:

A. product definition (including specifications)

B. engineering approach and risks

C. environmental and safety tests

D. project schedule establishing milestones

E. engineering cost to completion based on ROI

F. customer site testing plan

G. sales projections, competitive analysis, product promotion

H. manufacturing cost and plan summary

I. project team—engineering, marketing, manufacturing, product support. The manufacturing, marketing, and product support members should be actively involved in the project from the beginning and must be held accountable for their part of the project. All members of the team should participate in the development of the business plan and they should "sign up" indicating their commitment to complete the project on time within the budget.

3.2 Project Scheduling

Project schedules should be made out by the responsible person for all tasks including manufacturing, marketing, and product support responsibilities. Effective project management requires monitoring of the progress of the project against the schedule and early detection of problems.

3.3 Project Meetings

Project meetings should be held on a regular basis to facilitate communications between team members, resolve problems, and to recognize progress. It is also extremely important that the project manager regularly meet with team members on a one-to-one basis to stay on top of developments and demonstrate his support for the team member.

3.4 Manufacturing Plan
A detailed manufacturing plan should be developed as early in the project as possible. The plan should address the following:

A. Make or buy decisions for components and assemblies

B. Test philosophies and plan—built in, special fixtures, outside service

C. Capital equipment/Automation

D. Manpower

E. Floor space

F. Environmental test

G. Manufacturing Introduction Plan

3.5 Parts and Vendor Control
A preferred parts manual and preferred vendors should be developed at the start-up of the company. Parts and vendor proliferation are major sources of cost escalation and quality deterioration. These sources should be used on all new projects if at all possible.

3.6 Tooling Plan
The level of tooling chosen depends on production volumes. If there is doubt about the accuracy of the sales forecast, it may be better to choose a lower cost tooling alternative to minimize the cost and risks associated with changes to high volume tooling. If sales volumes exceed expectations, higher volume tooling can be developed to reduce manufacturing cost. This is always a difficult tradeoff since parts cost and quality can be related to the kind of tooling developed.

3.7 Reliability Goals
Reliability goals and a quality assurance plan should be developed for the product. Consultants are available to calculate the reliability of a design based on military handbook guidelines and failure rates. A calculated failure rate will give you a feeling for the expected failure rate that the product will experience in the field. That knowledge will help you estimate warranty costs and product support requirements.

3.8   Project Costing

A project costing report should be available from accounting on a monthly basis to track expenses against budget. Project members should be educated on the importance of accurate job costing for the project. This is important not only for the current project, but also for estimating costs for new projects.

3.9   Project Documentation

A strong argument could be made that the only reason for a research and development project is to produce documentation for manufacturing. Do not skimp in this area. There are many good CAD packages available for printed circuit design and drafting that greatly improve the documentation process for a project. There may not be significant savings in the initial design process, but changes and record keeping are greatly expedited.

3.10 Product Support Plan

A support plan that addresses repair, warranty, and response time should be developed for each new product. The plan should be written by the product support person that will be responsible for the product after it is released into manufacturing. Self-test, remote diagnostics, and customer documentation should be included in the plan.

3.11 Prototype Acceptance

A product is ready to be released to manufacturing after it has met all specifications, completed all environmental tests, successfully completed beta site tests, and is fully documented for manufacture. The product should be signed off at a team meeting which initiates the start of the manufacturing process. If the program is well executed with all team members committed to the success of the product, this meeting is a formality and there should be no surprises.

4. Product Introduction

A product introduction plan should be developed by marketing as an integral part of the product development cycle.

Components of Manufacturing Information Systems

Manufacturing information systems consist of an integrated group of software modules that address manufacturing, marketing/sales, financial, and management reporting needs of modern manufacturing companies. The simplest systems that usually run on personal computers comprise general ledger and manufacturing resource planning (MRP) capability that can handle a limited number of components in the inventory item master and schedule material and jobs in a limited way. The most robust versions—which require large minicomputer or mainframe capability—can be interfaced to various automated manufacturing equipment to form the core of a computer integrated manufacturing (CIM) system. This level of sophistication is currently found only at the largest companies.

At the time this book was written, there was a large array of accounting software packages available to automate the financial reporting of most businesses. The ability to handle complex manufacturing operations for planning and reporting differentiates the products offered by vendors of integrated MRP systems. You should concentrate your efforts in this area when evaluating potential candidates for your company. A high-end product should contain the capabilities listed below. The systems that run on personal computers will have fewer capabilities and less flexibility in manufacturing planning. Any system that you consider should be on-line transaction based, which provides you a real-time view of your company's daily operations.

The outline which follows provides an overview of the capabilities typically found in modern integrated general ledger/MRP software packages, and can be used as a guide to selecting an appropriate software package for your company.

1. Manufacturing
   The manufacturing module of the software package is the most complex part of the information system package and will dictate the hardware requirements of your system.

   Material Control

   ▼ Accomodates on-line transactions of stockroom issues and receipts, finished goods, work-in-progress (WIP), receiving and inspection, and customer shipments.

   ▼ Provides extensive reporting of obsolete and excess inventory and ABC part analysis.

   ▼ Maintains transaction log for audit trail of all inventory and standard cost changes.

   ▼ Cycle counts both stores and WIP locations based on ABC analysis and counting frequency.

   ▼ Allocates material to work orders and generates kit lists with shortage reporting.

   ▼ Prints purchase orders and requisitions as well as monitors delivery status and parts lead times.

   ▼ Provides vendor quality control monitoring.

   Engineering Data Control

   ▼ Maintains assembly–component relationships for up to 25 levels on bills of material for planning.

   ▼ Provides bill maintenance capabilities and access to costed, indented, and summarized bills of material.

   ▼ Provides "where used" reports for components.

   ▼ Implements engineering changes and revision labels through effectivity dates and historical reporting.

   ▼ Provides routing control for parts and product groups.

   ▼ Maintains work center set-up and labor rates, overhead rates, and capacity.

   ▼ Provides complete cost "roll up" capability for development of standard costs and variances, complete cost breakdowns for all parts including material, labor, and multiple overhead factors.

Master Planning

▼ Provides strategic planning capability, including master production scheduling, simulation of future delivery of products and parts, and options forecasting.

▼ Interfaces with order processing and service modules to include spare parts demand as part of parts forecast.

Material and Capacity Planning

▼ Uses master production schedule to determine material and labor capacity requirements in a user-defined planning horizon.

▼ Filters output of MRP to provide exception reporting on a user-defined basis.

Purchasing

▼ Accommodates purchase orders for regular inventory, expensed, and subcontracted items.

▼ Processes purchase orders for multiple parts and delivery dates, and shows commitments by date, vendor, or part number.

▼ Analyzes vendor performance, maintains approved vendor list, accommodates blanket purchase agreements, and projects future requirements for vendor price negotiations.

Shop Floor Control

▼ Tracks all information on production work orders, including activity, material and labor cost, overhead, outside processing, and job status.

▼ Provides work order scheduling.

▼ Provides reports by employee, operation, work center, work order, and scrap by fault code.

Cost Accounting

▼ Produces cost reports on open and closed work orders, purchased part variances, inventory, and work-in-progress,

▼ Maintains audit trail of material and labor transactions.

2. Order Management

The order management module is responsible for tracking of customer orders for the MRP system, maintaining accounts receivable and billings, and providing sales analysis and forecasting capability.

Customer Information

▼ Maintains customer bookings and sales, credit, and accounts receivable information.

▼ Allows flexible groupings of customers for pricing, reporting, and general ledger purposes.

Quotation Processing

▼ Provides on-line product availability "look-up" and configuration.

▼ Generates quotes or bids for prospective customers without affecting bookings or sales, and provides open quote reports for sales force follow-up.

Sales Order Processing

▼ Provides data entry screen with on-line product availability, product configuration, and credit check.

▼ Allows flexible pricing, including trade and prompt payment discounts.

▼ Provides acknowledgment and delivery information.

Shipping and Invoicing

▼ Checks customer credit at time of shipment, prints shipping documents, and tracks serial numbers.

▼ Supports several types of invoices, including proforma, miscellaneous charges or products, and foreign currency.

Accounts Receivable

▼ Reports summary and detail aging, and handles credit and customer debit memos for returns, adjustments, and disputed amounts.

▼ Provides detailed customer statements weekly, monthly, or quarterly.

▼ Accepts and tracks customer deposits and allows full or partial customer payments to be applied to any open item.

Sales Analysis and Forecasting

▼ Reports bookings, sales,and gross margin information by sales category, customer, territory, sales agent, product grouping, and product.

▼ Collects shipment quantities, shipment values, and serial number, and captures monthly bookings and sales data by product.

▼ Provides sales forecasting by product and customer, and relates the forecast to the current quotation data.

3. General Ledger

The general ledger provides all of the accounting information for the company plus budgeting for financial control.

Accounts Payable

▼ Creates vouchers from vendor invoices, calculates due dates and discounts, displays variances between standard, quoted, and invoiced costs, and matches purchase orders with receivers and invoices.

▼ Generates cash requirements reports and writes checks for payment.

Financial Reporting

▼ Generates month-end closing income statements and balance sheets with comparisons to previous periods and budgets.

▼ Summarizes project expenditures by account and permits reporting by cost center or department.

▼ Provides detailed daily transaction journals for audit purposes.

▼ Provides consolidation of budget and actual company data.

4. Report Writer

Any system you consider should have a report writer which allows your staff to easily obtain custom reports on the information contained in the data base. This feature can significantly improve productivity since the system users can tailor reports to meet their specific needs. Some vendors have developed a series of video display terminal reports that provide many common analysis tools without waiting for hard copy reports that are typically generated by the system.

The system should easily export data to the most popular spreadsheets—such as Lotus 1-2-3, Microsoft Excel, and Borland Quattro—so that your staff can develop their own "what-if" scenarios. It should also handle foreign currency transactions if your company will be doing a lot of international business.

Deming's Fourteen Management Principles

Start-ups are often so totally preoccupied with getting a product out the door that fundamental management practices are often set aside until "the company gets big enough to afford the luxury." It's better to start with the best management techniques in the beginning to establish a quality culture within your company. The following list, based upon the work of Dr. W. Edwards Deming, can serve as an excellent set of management guidelines regardless of the size of your company.

1. Create a constancy of purpose towards improving products and services and allocate resources for long-range needs rather than short-term profitability.

2. Plan for economic stability by refusing to allow commonly accepted levels of delays, mistakes, and defective materials and workmanship.

3. Don't depend on mass inspection; require statistical evidence of built-in quality in both manufacturing and purchasing functions.

4. Reduce the number of suppliers for the same item. Eliminate those that do not qualify with statistical evidence of quality. End the practice of awarding business solely on the basis of price.

5. Continually search for problems in the system so you can constantly improve processes.

6. Institute modern methods of training to make better use of all of your employees.

7. Focus supervision on helping people do a better job. Ensure that immediate action is taken on reports of defects, maintenance requirements, poor tools, inadequate operating definitions, or other conditions detrimental to quality.

8. Encourage effective, two-way communication and other means to end fear throughout your organization and help people work more productively.

9. Break down barriers between departments by encouraging problem solving through teamwork. Combine the efforts of people from different areas such as research, design, sales, and production.

10. Don't use numerical goals, posters, or slogans that ask for new levels of productivity without providing methods.

11. Use statistical methods to continue improvement of quality and productivity. Eliminate work standards that prescribe numerical quotas.

12. Remove all barriers that inhibit a worker's right to pride of workmanship.

13. Institute a vigorous program of education and retraining to keep up with changes in materials, methods, product design, and machinery.

14. Clearly define top management's permanent commitment to quality and productivity as well as its obligation to implement all of these principles.

Glossary

Alpha testing: the testing, usually at the work site of a typical customer, of the initial version or prototype of a product.

Angel: a private individual from the outside who invests in a start-up venture.

Beta testing: the testing, usually at the work site of a typical customer, of a pre-production version or prototype of a product incorporating changes made as a result of the alpha testing process.

Bill of materials (BOM): a detailed list of the parts and other items used to manufacture one unit of a certain product.

Business plan: a written document describing how a company will be organized and managed, the products it will produce, and the expected financial results for a certain period of time.

CEO: acronym for "chief executive officer," the highest ranking executive within a company.

CFO: acronym for "chief financial officer," the highest ranking executive responsible for the financial aspects of a company. This person often reports directly to the CEO.

Computer integrated manufacturing (CIM): the control and interlinking of all elements and stations of a manufacturing process by computers.

Design-in: when a product is selected by a manufacturer as a component in an item they produce.

DFMA: acronym for "design for manufacturability and assembly," which is the process by which considerations of how well and economically a product can be produced are incorporated into the design of the product.

ECN: acronym for "engineering change notice," which is written documentation of every engineering change, even as minor as substituting a single part, made to a product design.

IPO: acronym for "initial public offering," which is the first time a company's stock is offered for sale to the public.

JIT: acronym for "just in time," a technique for minimizing inventory costs by scheduling frequent deliveries of parts and materials for the times when they are needed instead of keeping extensive supplies of them on hand.

Mission statement: a concise definition of a company that tells the company's products, technologies, customers, and intended applications.

MPS: acronym for "master production schedule," a detailed plan describing how many units of a product will be manufactured, the time periods in which such units will be made, and the parts and other supplies that will be needed.

MRP: acronym for "manufacturing resource planning," a technique using highly integrated software to provide management with information on the resources needed in a manufacturing company, the availability of such resources, and how to most efficiently manage them.

Objective: a quantitative statement of a goal to be achieved by a specified time.

OEM: acronym for "original equipment manufacturer," a company that manufactures items for sale to end users.

Price-earnings ratio: the ratio of the price of a company's stock to its after-tax profit.

Seed funding: the very first financing received by a new company at its inception.

Strategy: a plan describing one means by which a corporate objective will be achieved.

Tactic: a set of specific actions used to carry out a strategy.

TQC: acronym for "total quality control," a concept of increasing quality through continuous improvements and refinements in all aspects of a company's business.

VAR: acronym for "value-added reseller," a firm that buys individual components (such as those making up a computer system) with a specific application in mind and "adds value" to the components, usually through custom software or integration and configuration of the individual components.

WIP: acronym for "work in progress," inventory of a higher level than raw material but lower than finished products.

Bibliography

Marketing and Strategy

Davidow, William H., *Marketing High Technology: An Insider's View.* New York: The Free Press, 1986.

Davidow, William H. and Uttal, Bro, *Total Customer Service—The Ultimate Weapon.* New York: Harper & Row, 1989.

McKenna, Regis, *The Regis Touch.* Reading, Massachusetts: Addison Wesley, 1985.

Ries, Al and Trout, Jack, *Positioning: The Battle for Your Mind.* New York: Warner Books, 1981.

Ries, Al and Trout, Jack, *Marketing Warfare.* New York: McGraw-Hill, 1986.

High-Tech Entrepreneurship

Bell, C. Gordon with McNamara, John E., *High-Tech Ventures: The Guide to Entrepreneurial Success.* Reading, Massachusetts: Addison-Wesley, 1991.

Stolz, William J., *STARTUP: An Entrepreneur's Guide to Launching and Managing a New Venture.* Rochester, NY: Rock Beach Press, 1988.

Trudel, John D., *High Tech with Low Risk.* LaGrande, Oregon: Regional Services Institute of Eastern Oregon State College, 1990.

Quality

Crosby, Phillip B., *Quality is Free: The Art of Making Quality Certain.* New York: McGraw-Hill, 1979.

Crosby, Phillip B., *Quality Without Tears—The Art of Hassle-Free Management.* New York: McGraw-Hill, 1989.

Grant, Eugene L. and Leavenworth, Richard F., *Statistical Quality Control.* New York: McGraw-Hill, 1988.

Finance

Seiler, Robert E., *Principles of Accounting: A Managerial Approach.* Indianapolis: Charles E. Merrill, 1967.

Weston, J. Fred and Brigham, Eugene F., *Essential of Managerial Finance.* New York: Holt, Rinehart and Winston, 1974.

General Management

Drucker, Peter, *The Effective Executive.* New York: Harper-Collins, 1967.

Drucker, Peter, *Managing for Results.* Harper-Collins, 1986.

Drucker, Peter, *The Practice of Management.* Harper-Collins, 1986.

International Sales

Wells, L. Fargo and Dulat, Karin B., *Exporting: From Start to Finance.* Blue Ridge Summit, PA: Liberty House Books, 1989.

Following are some reference books that can help you evaluate foreign market opportunities:

The Export Guide to Europe, published in association with Worldwide Intelligence by Gale Research Co., 1986. This reference provides information on market size, appointing agents or distributors, investment incentives, trade fairs, standards approval (e.g., electromagnetic interference standards), payment practices, debt collection, and major buyers by product interest.

Kompass, distributed in the U.S. by Croner Publications, Inc., New York, NY. This reference, published in various European countries, contains detailed cross-references from product to companies that sell or manufacture the product. The reference also includes company general and financial information. It can help you find representatives or distributors that sell compatible product lines.

Exporter's Encyclopedia, published by Dun's Marketing Services, and updated annually. This guide is comprehensive, covering world markets from Argentina to Zimbabwe. The guide contains information on a country's communications, transportation, information sources, and markets. Key U.S. and foreign contacts are listed. The guide also includes information on export documentation, trade regulations, and business travel.

Predicasts F&S This publication contains market data by Standard Industrial Classification (SIC) code by country. It also reveals market trends. The SIC code categories can provide fine detail on market segments.

Index

Accounting services, 41–42, 177–178
Activity ratios, 187–188
Advertising, 104–105
Apple Computer:
    founding of, 7
    just in time (JIT) manufacturing
        by, 156
    lawsuit against by Xerox, 48
    licensing STAR technology, 8
    "Lisa" computer, 8
    marketing and sales expenses
        of, 180
    price/performance ratio
        of, 109–110
    problems due to vendor quality
        control, 164
    quality testing by, 173
    Regis McKenna and, 13
Applications development, 107
Average collection period ratio, 187–188

Balance sheet, 181–183
Boards of directors, 40–41
Brand names, 37
Bubble memories, 145–146
Build or buy decisions, 159–160
Business plan:
    assistance in developing, 32
    outline for, 211–214
    preparing, 30–32

Campbell Scientific, 172
Cash flow, 194–196
"Cashing in," 77–80
Catalogs, 122
Cellular Data, Inc. (CDI), 74
Center for International Trade
        Development (CITD), 130
Compaq Computer:
    management information system
        at, 203
    positioning of, 28
Competitive analysis, 99–100
Computer integrated manufacturing
        (CIM), 168
Control Data Corp. (CDC), 145
Copyrights, 47–49
Corporate culture, 94–96
Cost control, 181
Current ratio, 184
Customer training, 140–141

Debt to total assets ratio, 185–186
Dell Computer, 109
Designing for manufacturability,
        153–154
Direct mail, 105
Direct sales force, 118–119
Distribution partners, relations with,
        126–127
Documentation, 141–142
Domain Technology, 52
Domino's Pizza, 176
DSP Group, 130

EIP Microwave, 25–26
Employee benefits:
    financial compensation, 88–89
    medical, 89–90
    retirement plans, 90–92
    stock options, 87–88
    time off, 92–93
Employment interviews, 84
Executive recruiters, 83

Ferretec, Inc., 32, 103
Financial reporting, 179–187
Financial statements, 191
Fixed charge coverage ratio, 186
Funding sources:
    banks, 64–65
    customers, 61–62
    friends and relatives, 61
    government contracts, 62
    government enterprise funds, 63
    individuals, 61
    research and development
        partnerships, 63
    Small Business Administration, 63
    universities, 64
    venture capital, 61, 65–72

Handar, Inc., 40, 146, 173
Hewlett-Packard (H-P):
    calculator pricing by, 110
    planning at, 17
    product development at, 8, 145–146
    sale of technology to Trimble
        Navigation, 6
Hypres Corp., 112

ILX Lightwave, 26
Incorporating, 37–42
International selling, 127–137
Inventory turnover ratio, 187

Joint manufacturing venture, 73–75

Kalok Corporation, 59–60

Leverage ratios, 185–187
Lightwave Electronics, 2
Liquidity ratios, 183–184

Management information systems
    (MIS), 237–242
Manufacturing resource planning
    (MRP), 165–167
Marketing research, 15–17, 99
Master production schedule (MPS), 158
McKenna, Regis, 13, 146
Medical benefits, 89–90
Medical Optics, 17
Miniscribe, 52
Mission statement, 18–19
Mortensen, Bob, 2

Naming a company, 35–37
New product development, 100–101,
    142, 143–154
Newport Corp., 99
NeXT Corp., 112

Objectives, corporate, 19–20, 22–25
Original equipment manufacturers
    (OEMs), 51–55

Palo Alto Research Center (PARC), 8
Partnerships and joint ventures, 73–75
Patents, 43–47
Performance evaluations, 85–86
Personnel administration, 85–86
Photon, Inc., 97–98
Pricing, 108–111
Product positioning, 27–29
Profit and loss (P&L) statement, 179–181
Profit margin on sales ratio, 189
Profitability ratios, 189–190
Promotion techniques:
    advertising, 104–105
    direct mail, 105

literature, 106
newsletters, 105–106
publicity, 104
seminars,105
trade shows, 105
videotapes, 106
word-of-mouth, 104
Publicity, 104

Quanta-Ray, 8–9, 53–54, 100, 116
Quantum Design, 28–29
Quick ratio, 184

Recruiting employees, 82–85
Research and development:
expenses by company, 147–149
partnerships, 63
through technology licensing, 149
Retirement plans, 90–92
Return on investment (ROI) ratio, 217
Return on total assets ratio, 190

Sales forecasting, 102–104
Sales representatives, 120, 123–125
Sales training, 108, 114
Saxpy Computer, 28
Selling channels:
catalogs, 122
direct mail, 121–122
direct sales force, 118–119
retail outlets, 121
sales representatives, 120, 123–125
stocking distributors, 120–121
telemarketing, 119–120
Small Business Administration (SBA), 63
Small Business Innovative Research
(SBIR) program, 62, 150–151
Strategic planning, 20–22
Stock options, 87–88
Stocking distributors, 120–121
Subchapter-S corporation, 37
Sun Microsystems, 18, 146–147, 149, 177

Tactical planning, 22
Tandem Computer:
corporate culture of, 95
employment interview
techniques, 84
product development by, 145
strategies of, 4, 23–25
values of early employees, 9
Technology licensing, 149
Termination of employees, 86
Time off for employees, 92–93
Times interest earned ratio, 186
Total asset turnover ratio, 188
Total quality control (TQC), 171–172
Trade secrets, 47
Trademarks, 49–50
Trial marketing, 102
Trilogy Corporation, 10–11
Trimble, Charles, 3–4, 9
Trimble Navigation, 6, 88, 160, 180, 203
TSI, Inc., 141, 150–151

Valuation of a company, 75–77
Value perceived by customers, 108–109
Value-added resellers (VARs), 122–123
Venture capital:
advantages of, 67–68
disadvantages of, 68–69
negotiating for, 71–72
stages of, 66
Versatec, Inc., 86

Xerox:
lawsuit against Apple, 48
Palo Alto Research Center
(PARC), 8
STAR system, 8

3Com Corporation, 11, 146